# Ancient Destiny Volume III

## Generations



ALBERT LYNN CLARK

**Books by Albert Clark**

**Fiction**

**The Price of Peace (1987, 2012)**

**Ancient Destiny (Volume I) SHIP (2012)**

**Ancient Destiny (Volume II) Escape from Earth (2012)**

**Ancient Destiny (Volume III) Ancestors in Space (2012)**

**Non-Fiction**

**Computer Supported Network Analysis System (1981) CSNAS**

**Guide to Acquisition Logistics (1985)**

**Various US Air Force Manuals**

ISBN-13: 978-1478323211

ISBN-10: 14783232

# Contents

# DEDICATION

To my wife, Carolyn Clark who humored me when I spent hours at the computer writing, searching the internet, reading my email.

# ACKNOWLEDGMENTS

Thank you to the education I received. Thank you to all the science fiction authors out there. Thank you for my 33 first cousins who spent days and weeks discussing strange things about science, possible futures, and sharing their books.

| Years | Who | Where | What |
|---|---|---|---|
| 5500BC | | | Copper Smelting |
| 3500BC | | | Bronze & Cuneiform writing. |
| 570-495BC | Pythagoras | Greece | Mathematics |
| 427 - 327BC | Plato | Greece | academy for 876 years, tried to tell people about Atlantis |
| 384 - 322BC | Aristotle | Greece | Student of Plato and known for many scientific principles |
| 356 - 323BC | Alexander the Great | Greece | Greek who conquered most of the known world. |
| 287-212BC | Archimedes | Greece | specific gravity |
| 90-168AD | Ptolemy | Greek in Egypt | Founded Museum of Alexandria |
| 1027 | Mayan | South America | epidemic |
| 1050 | | Norway | Oslo founded |
| 1215 | King John | England | Magna Carta |
| 1473-1543 | Copernicus | Poland | Earth circles the sun instead of vice versa. |
| 1233-1559 | | | Spanish Inquisition |
| 1532 | | Inca | Spanish destroyed Inca civilization |
| 1514 - 1564 | Andreas Vesalius | Belgium | Human Anatomy |
| 1491 - 1547 | King Henry VIII | England | Break from Catholic Church in 1534 ended inquisitions |
| 1544 - 1603 | William Gilbert | England | Electricty, attraction, force, magnetic pole |
| 1564 - 1642 | Galileo | Italy | telescope |
| 1642 - 1727 | Sir Issac Newton | England | English mathematician and scientist who invented differential calculus and formulated |

| | | | |
|---|---|---|---|
| | | | the theories of universal gravitation, terrestrial mechanics, and color. His treatise on gravitation, presented in Principia Mathematica. |
| 1727 | J.H. Schulze | Germany | Photography |
| 1698 | Edward Somerset | England | steam engine |
| 1755 - 1804 | Alexander Hamilton | USA | Proved that lightening was electricity. Build huge industrial complex near Albany, NY that pushed US into industrial age. |
| 1807 | Robert Fulton | USA | Clermont made 150 mi trip from NYC to Albany in 32 hours |
| 1819 | | | First steam ship crossing of Atlantic |
| 1838 | | | Steam ship crossed Atlantic solely on steam power |
| 1866 | Jean Lenoir | Belgium | Internal Combustion Engine |
| 1892 | Stephenson | England | Steam Locomotive |
| 1847 – 1922 | Alexander Graham Bell | USA | Telephone |
| 1828 – 1905 | Jules Verne | France | Author with ideas on-rockets, atomic sub, communications, etc. |
| 1863 – 1947 | Henry Ford | USA | Assembly Line, but many automotive inventions |
| 1879 – 1955 | Albert Einstein | USA | Physics and Math |
| 1908-91 | Bardeen, John, | USA | Transistor |
| 1912 – 1977 | Wernher von Braun | USA | Started in Germany, but is father of US rocket |

| | | | |
|---|---|---|---|
| | | | science. |
| 1957 | Multiple | Russia | Sputnik |
| 1960 | Multiple | USA | Computer |
| 1976 | Stephen G. Wozniak, and Steven Jobs | USA | Apple Computer |

# FOREWORD

# BACKGROUND IN SCIENCE

The history of Earth is not quite what is in the history books. First there was Mars. The sun was hotter back when dinosaurs roamed Earth and people lived on Mars. However, the sun had cooled over millions of years and the dinosaurs had died out. The Martians had already sent exploratory missions to Earth and had a few small colonies there. The Martians discovered a large interstellar body that was going to impact Mars as a catastrophic life ending event. In response they built spaceships more advanced than what we have today and evacuated tens of thousands to Earth.

There is a Rift Valley on Mars 4,000 kilometers long or about 2,500 miles long and 120 miles wide that appears to have been formed by an interstellar impact.

http://en.wikipedia.org/wiki/Valles_Marineris If it was caused by impact, it would have been sufficient to blow away its atmosphere and end life on Mars. What little remains of Martian atmosphere is similar to our own, but very thin. There is ice at the Martian poles that could be converted to oxygen to breathe and hydrogen for fuel as well as just melted for water.

Phase two of this history was our history of Atlantis. Our history treats it as both a far fetched and possibly true story by Plato, and an ancient city-state that was destroyed by a volcano. Many scientists believe that Atlantis actually existed. Some said it was outside the Pillars of Hercules, in the Atlantic Ocean some said it was in the Mediterranean.

Some scientists actually name the Greek Island of Santorini as Atlantis. I have been there and have the photos. Today it has a village on a high cliff you can reach from the sea by donkey, walk, or take cable cars to the top. From the top you can see a large bay curving away from the village with a sometimes smoking remnant of a volcano at the distant point of the crescent shaped bay. There are frequently several cruise ships and other boats in the large bay. The story goes that the volcano exploded, destroying Atlantis and just leaving the volcano and the crescent shaped bay remaining.

The stories say that Atlantis had crystals on the mountain peaks that sent out power for airships to roam the earth. Descriptions of the actual city-state depict Atlantis as a great city with advanced technology with multiple circular canals around the city.

One story of Atlantis describes an army of 10,000 Atlanteans easily beating back hundreds of thousands of Arabs in Northern Africa and southern Europe. Every battle is a monumental victory with thousands of defenders dying with almost no losses on the Atlantean side. However, when Egypt and Greece were about to be overrun, the Atlanteans received word that something bad was happening in Atlantis and they all retreated home just before Atlantis was destroyed and forever lost. Does this mean that humans were on Earth when the Martians came and established Atlantis? Does this mean that maybe colonies from Atlantis were attempting to rebel against Atlantis rule? Did Atlantis retreat because of news of an impending disaster in Atlantis?

Or, was Plato simply describing a fictional Utopia that never existed. Plato said he was reading ancient documents. History says that 200 years after Plato, the Greeks built the library of Alexandria to hold all the great documents of the world, but the library was destroyed and all the documents were lost in a fire in 48 B.C. when Caesar decided to burn his fleet to prevent Achilles from communicating by sea. The burning of the library was accidentally caused by the burning of the fleet. That is only one story. Some say it was destroyed by Christians against paganism. Some say it was the Moslems around 900 A.D. I think we can assume the library existed. Did it hold the history of

Atlantis? Did it contain the science that Atlantis had? We'll never know. Edgar Cayce, a modern day profit, said we would find a library of Atlantis.

There is a whole area of pseudo science called Forbidden Archeology where there is evidence of very advanced, very old societies predating Egypt. There is the brass mechanical computer attributed to ancient Greece. It was found in the Mediterranean and encrusted. X-Rays reveal multiple gears to compute. Just get on the internet and search for "forbidden archeology". There is a coin discovered that appears to be 200,000 years old, about 195,000 years before coins. Long before the Egyptian civilization came to power, but do your own looking at forbidden archeology. Don't forget that sometimes archeology is spelled archaeology. Keep in mind ancient American Indians were in the Americas 12,000 years ago. The Mayans and Aztecs did not just suddenly appear as advanced races rivaling the Egyptians.

All of the above is evidence that at least some of the story could be true.

Just a note in my own history. When I started in first grade in grade school I was reading Popular Science and Popular Mechanics magazines and many novels while the other first graders were learning "See Tom Run". School taught that all airplanes had to have propellers and my teachers and classmates could not understand when I was building clay model jet aircraft while they had propellers on their clay airplanes. I was told I was wrong. I was taught that nothing could exist above the speed of sound, but I knew that Chuck Yeager had done just that in 1947 when I was only 3 years old. I read all about it when I was in first grade, so by the second grade I was arguing that Chuck Yeager had already done it and survived. I was taught that E=MC2. At the speed of light all material is changed to energy. Recently I have read that scientists have evidence that some particles have exceeded the speed of light. Star Trek has used Warp Drive for multiples of the speed of light.

# CHAPTER ONE

# SPLITTING UP

Sandy, Dawn, and Craig took SHIP back to Pangaea to consult with the rest of the crew. Enroute they discussed their next major mission. During the three weeks of their voyage they looked at what information they had and decided they could make no decisions without the rest of the crew.

When they arrived in orbit over Pangaea each of the other four female crew members were teleported to SHIP for the crew meeting on their next mission. Once they were all together around the dining table, the dining room was a combination dining room and library. The old fashioned library was filled with antique leather covered books in dark walnut wood glass enclosed shelves. The other walls were apparently wood paneled with an intricate sculpted ceiling with a large crystal chandelier over the long shiny dark wood table. There were three side chairs on each side and a captain's chair with arms at the head. When they ate, SHIP teleported whatever food each one wanted in front of each person perfectly cooked indistinguishable from real food on real expensive fine china with wine, cola, water, whatever the person wanted served in fine crystal glassware, the silverware was real and solid Sterling silver. The food was cooked and seasoned to match the highest expectations of whoever picked it. However, everything was manufactured by SHIP for each meal from its intensely packed subatomic structure, so everything was really a copy.

Craig started out by summarizing their history. "You know this history, but let me summarize before we address our next mission. SHIP teleported me here to SHIP, several years ago, educated me in the learning chair about the history of the ship and gave me the warning about the pending destruction of Earth by an interstellar body that had previously destroyed the Martian civilization. Prior to impact some of the Martians had escaped to Earth and created the Atlantean civilization."

"Once again, they had warning that Atlantis would be destroyed by nature and escaped to the stars to continue their civilization on other planets in other solar systems. When Atlantis was destroyed all the transmitted energy of Atlantis stopped, knocking Martians and humanity back to the Stone Age. They had a prediction of that same object that destroyed the Martian civilization was going to come back again, this time with a direct instead of glancing impact with Earth and completely destroying civilization and possibly all life on Earth. They left the ship behind to train the earthly survivors of Atlantis and other humans on Earth to escape into space once again. Dawn, my legal wife, and I actually visited an abandoned Martian command post under the polar ice while on our honeymoon on SHIP and saw, first hand, that this history was true. We witnessed the destruction of Mars on recorded media and saw what Mars was like before its destruction. The biggest impact was the loss of the Martian atmosphere which made Mars uninhabitable."

"Over the thousands of years since Atlantis was destroyed, SHIP had selected individuals to be trained here on SHIP. The various super scientists of Earth were mostly trained by SHIP to advance science. Some tried to advance science too fast and failed to be replaced by another individual. Some lost their minds by trying to absorb too much knowledge from SHIP."

"We know some of these: Aristotle, da Vinci, Edison, then close behind, Tesla, Henry Ford, and others. They were highly respected when alive and even more famous after their passing. SHIP cut some off in their prime because they were interested in personal profit, some because they pushed too hard. One example was Tesla trying to repeat the energy transmission used by Atlantis. If successful, electrical energy would be transmitted around the world just as radio and television are transmitted.

Everyone, and every mode of transportation would have had unlimited access to electrical power and zero pollution. It was too soon, Tesla failed and was discredited and SHIP cut him off, leaving Tesla to be viewed as a nutcase."

"I was the last selected by SHIP, but with a very limited time to get humanity into space away from Earth before its destruction. Then it turned out that even the Martians were fallible in predicting the return of the destructive interstellar object, rouge moon, or whatever it was. It was returning several years earlier than planned. I needed help, but Martian programming of SHIP only allowed one male captain and six females to come to the SHIP. This was the ideal mix for establishing a new colony, but was not really prepared to be polygamous. Dawn convinced me that it was necessary to get the scientific team needed. I selected those six and together, you and I, we managed to evacuate over one hundred forty thousand humans to three other planets in three separate solar systems."

"I was mainly the figurehead or leader, but it was you girls that actually made it happen. Each of you received specialized training in your fields to include all the historical knowledge of Mars and Atlantis as well as Earth in your area of specialization. One person would have overloaded their minds in trying to learn all these critical areas and been totally crazy if they lived. So the knowledge of creating interstellar ships and starting new colonies is shared by each of you learning just one part of the needed science."

"Now, Sandy, Dawn, and I have found more planets that were probably inhabited by the Martians that are now void of humanity. Sandy would you tell us your opinions of what we found." As they went around the table, Craig listened closely to each of his team of experts, but reflected on each of them.

Sandy was losing her dark California beach surfer's suntan. They had been bottled up on the ship most of the time for months now. She was still pretty with her sun bleached hair turning a shade of light brown instead of blond. She had a trim and slim body of someone that lived on the beaches of California. Every time they were teleported to SHIP, SHIP kept each of them in prime physical condition and returned them to superior health. Sandy's tan was fading because, on Earth, she tended to be overexposed to the sun so SHIP had allowed her tan to fade to a more reasonable level. SHIP even healed any past injuries or genetic disposition toward future problems. They would stay young as long as they teleported to and from SHIP. Any injuries would be healed with no trace when they teleported. Sandy would stay very close to her five foot five and one hundred twenty pounds with unlined skin forever. Sandy was very much in love with him and he with her, but he had added her as a crew member rather late. When he had the future of the human race in his hands he was cautious about his heart ruling his selection process. Their mental specialties had been primary.

"Okay. Keep in mind that I am the Zoologist and Veterinarian, not truly a scientist. We did not search the whole planet, only the area around a defunct city. It appeared that all of their metal just disintegrated with the remains of ground vehicles in the streets between collapsed buildings where the metal superstructure had disintegrated, no longer supporting the buildings. Apparently all animal life, at least in that area was gone causing carbon dioxide too low to even support algae in standing water. There were orchards near dead, probably for lack of carbon dioxide for photosynthesis. Dawn, your impressions."

Dawn was his first crew member from the earliest days. They had actually been married in front of her parents and the only legal wife. She was very attractive with a roundish face, light complexion, dark brown hair, and blue eyes. She was five foot four, one hundred twenty pounds and very physically fit. Not heavily muscled, but still firm. She was perfect in every way physically, but it had been a marriage of convenience because she wanted to go to space and he needed her knowledge. He liked her very much; they were very comfortable with each other like an old married couple and really had been together for a number of years. But it was the mission that had come before love. That was mutual and the marriage had been to satisfy her parents when they were going to be living together for years. She wanted to be the supreme astral navigation expert in the world and now she was. She never really hoped to go into space, but had planned the courses between the planets to get their colony ships to their new planets. She was the first human female to the Moon and Mars. They respected each other and were very close. Craig wondered if there was any love for him on her part. It was her idea originally to add more crew members. SHIP had confirmed that it was the Martian plan for the establishment of colonies: six women to each man. Women made better space voyagers and just as smart as the men and they could produce children to populate a new planet quicker with a higher proportion of women to men.

"Something caused their metals which may have been the Impervium to just disintegrate and maybe that same thing vaporized all animal life. We don't know that. It could have been an act of war using some form of energy to do the damage. It could have been the Impervium has a natural decay that happens all at once destroying the metal."

"We don't know whether this happened before or after animal life disappeared. It could be that the human or Martian or other life forms evacuated taking all their animals with them before the metal disintegrated. Or, maybe they evacuated after the metal disintegrated. We have no idea how old that city was or when whatever happened. Craig?"

"That was my impression too. If Impervium has a natural decay point, then we need to rethink our total dependence upon it. It was perfect for building light weight spaceships and establishing our colonies on Pangaea, Pyramid, and Vulcan, but if that life is short, we need to modify our whole plans to not be dependent upon Impervium. All our petroleum pipelines, drilling rigs, tools, even the nuclear reactors on the ships could suddenly disintegrate. Or, it may take many thousands of years for it to disintegrate. If it was an act of war, then that means that we have a threat from some other intelligent race in our galaxy."

"I think we need to go back and spend some time studying those deserted planets to see if the destruction was planet wide, what might have caused it, some clue as to what causes Impervium to decay. Was the decay sudden or over thousands of years? Let's go around the table and get everyone's opinion on what we need to do. Estera, your comments, please."

Estera was a petite little thing of five foot one and maybe ninety-five pounds, if that. Everything was in the right place with a tiny waist and slim hips. Her hair was darker brown and always short with green eyes. SHIP had recommended her as a biologist, but she looked more like a librarian with large thick "coke bottle" glasses when he met her the first time. SHIP had recommended her for two reasons. She was very smart and studying biology at Skidmore College in Saratoga Springs New York. He needed someone interested in biology. The second reason was that her father was the chairman of the federal House Ways and Means Committee and had political power to help him. She believed his story about the short future of Earth as did her father. Her father had practically blackmailed him to take her as an unofficial wife and crew member to get her off of Earth before its destruction. They liked each other well enough and had sex from time to time as he was the only male for months at a time, but mostly they were crew members. When she first transported to SHIP she thought she was going blind because she could not see with her glasses until she discovered that SHIP had corrected her eyesight to twenty ten during the teleportation.

"I agree with you Craig, it appears our next major mission is to investigate those other planets in more detail. I want to look at the plant material you brought back. However, I need to stay here on Pangaea and get our gardens and orchards growing well.

We can use the grasses for crude bread, but wheat, oats, barley, and other Earth grain grasses are much better adapted for farming. I would like to study the plant life here on Pangaea a lot further too. I don't see how I can help on these new planets. Craig, I don't want you to think I am abandoning you, but you know I agreed to join the crew to save my own life and because my father forced you to take me on to get his government support. I think I have helped, but my work is here. Pyramid was readied for us by the Martians when they came there in history, planting orchards, grains, even starting mines for metals and preparing key roads and stone housing. I don't care about Vulcan, but they will survive. I think Pangaea needs my help more than you do. If we are going to Craig's left, then it's your turn, Karen."

Karen was half Apache Indian and had black hair that sleekly stretched down her back past her shoulder blades half way to her waist and nearly black eyes. Her face was all angles with high cheek bones and strong chin. She was five foot seven or eight and one hundred forty-five pounds of pure muscle. She was his survivalist brought along to protect them from unknown hazards on new planets. She was an incredible shot with both bow and rifle. Her instincts were superior and seemed to sense danger where no one else saw it until she had taken care of it. Ship had recommended her and he had met up with her while she was walking in the Colorado Mountains alone and had tagged along gradually introducing her to his SHIP given strength and impossible technology until he convinced her to join the crew. Were they in love? He didn't think so. She was like am Amazon warrior and sexy as hell. He was very attracted to her and vice versa, but probably could never consider tender love. Craig thought, "I am only five foot ten and one hundred sixty pounds. Whereas Karen's muscles stood out whenever she just stood looking out for danger, Craig was not muscled like that. Oh he had muscles which were augmented by SHIP, but his muscles were not body builder type. Her muscles were not like a body builder either, but that of a hunter, which she was, sinewy. Craig was unremarkable, with thick brown hair and brown eyes that turned hazel when relaxed."

"Okay, I agree with Craig, we need to study the new planets. I stayed behind this last time to make sure that the Pangaea colonists could handle security from the natural predators here, but they can handle it without me now. I am not needed here. I think I should go into the learning chair and learn archaeology to see if we can unearth some history there. I did not saturate my mind in the learning chair before as SHIP could not teach me a whole lot about hunting and fishing and survival that I did not only know, but in many ways knew more than SHIP. The only thing I learned from SHIP was training methods which were largely mine to start with. I should be able to learn the detailed history of Mars, their written language, and whatever techniques I need without threat to my health. Dawn, I see you are antsy to jump in here."

"I also want to learn archaeology. SHIP could not teach me interstellar navigation; that was mainly me getting some electronic observations, measurements, math pointers and instant calculations of my formulas from SHIP. I think I have taught SHIP more than SHIP taught me. I could take some additional training without hurting my mind also. I just wanted to get that in and two of us cooperating should be able to do more than one alone. With my education in physics, I think SHIP could safely teach me additional physics to help determine what happened to the metals on these new planets. Jill, my old roommate from college, I think you are next."

Jill Leary was his real scientist. She had been majoring in petro chemical engineering and had been trained by SHIP in that field and in metallurgy, mining, and mineralogy. She was five foot four, auburn hair, brown eyes, and built. Even her complexion had that reddish glow, not a freckled redhead, but her skin just picked up on her auburn hair. She had regained much of her tan from working outdoors on Pangaea. She was a science fiction fan. She was Dawn's sorority sister and when Craig had pursued Dawn as his first shipmate she had been there more than once and heavily flirted with Craig. She was the proverbial tease, not promiscuous, but definitely a tease. She was sexy and knew it, but was more intent on her studies and career than men. Jill and Craig were still sexually attracted to each other and she was an enthusiastic partner.

"Craig, don't take this the wrong way. I would love to go with you to study the disintegrated metal, but with that news I am really needed here. Rather than just making everything from Impervium I need to create a factory to make plastics from the abundant oil here on Pangaea and we need to modify the metal works to produce other metals such as old fashioned steel. We were talking hundreds of miles of Impervium pipeline for petroleum transport around the continent. Now I am leaning to plastics and conventional metals that we know more about. Maybe the Martians guided us into Impervium because it was so super strong and light for our spaceships, but maybe it is not so great in the long term. Besides, plastic is easy to make with an abundance of fossil fuels available. Even if it were not for the flaw in Impervium, there is so much to do here. We don't really know that much about climate and need to get the oil fields and pipelines in place as quickly as possible. There is plenty of wood to build with, but is not a good material for building large constructions like refineries, factories, and so on. I really need to stay here. I can look at the dust from the Impervium in the lab, but don't really see the value in me studying those new dead planets. I will miss our nights together, but I need to stay here."

"I hear you Jill. I agree with you staying here and I will miss you too. Estera, you need to stay here and work on the fields and orchards. Okay, Sandy, you were with us on these new planets, what do you think?"

"I have already said we need to study the new planets and try to find out what happened there, but I need to stay here and get the Earth animal population growing. We have to be very careful on the test tube cows, pigs, chickens and other Earth animals. They are test tube animals. They will have to be incubated, then fed until they can be turned out into the fields to eat on their own. We brought other experts, but I am the only one with SHIP's training in this, so I need to stay here. You know how I feel about you, but I will trust Dawn to watch out for you and Karen to protect you. I need to stay"

"I agree, we are just letting Vulcan survive however they can, Pyramid has everything they need, but Pangaea has only the Pangaea buffalo for meat. We need to get Earth animals going, and I will miss you too. Shelby, I didn't mean to skip over you. What do you have to say?"

Craig picked Shelby Denton as a chemistry major but had long admired her as a standout from the rest of the cheerleaders. Where the other cheerleaders tried to act sexy when they did a high kick, Shelby kicked higher looking more like a dancer and looked graceful instead of show off. Typical of cheerleaders in this age of tossing girls in the air, she was only five foot three and weighed maybe one hundred five pounds. She was very into science fiction and did not take a lot of convincing to become one of Craig's crew members. She was quiet, but efficient. She had run various tests throughout their explorations on the water and plants to make sure they weren't poison or polluted. She tested the quality of the oil they had found.

"I want to stay here and work with Jill on developing plastic products. Yes, that is chemistry as much as manufacturing. I will test the samples you brought back with you on this last trip and you can collect other samples for me to test, but I would expect to find nothing except low carbon dioxide absorption in the water and plants. Plastics from the coal and petroleum may be a lifesaver. I learned some new things from SHIP about chemistry which includes better plastics. It was not so important before because we were using Impervium metal for everything. Impervium was impervious to heat and impact and was as light as Styrofoam while being stronger than steel ten times thicker. The plastic is also easier to manufacture and use. I can make fabric for clothing and upholstery as well as pipe, structural components, kitchen sinks, countertops, you name it. The plastic will be heavier than Impervium and far less strong, but can be made nearly fireproof. Nearly anyone can learn to use plastics. Yes, I need to stay here for now."

"Well, I guess that is Dawn, Karen, and I going to the new planets to try to find out what happened. The four staying here, remember that SHIP won't be here to teleport you out of danger or heal your wounds. You get a sunburn, you will have to suffer. We won't be gone more than a few months at most. If any of you get seriously hurt, I expect all of you to do everything in your combined power to keep them alive until we return. Go ahead with any medical treatment you can get. SHIP can cure anything as long as you are alive and your mind is still intact."

# CHAPTER TWO

# NAVIGATION MADE CLEAR

Those staying on Pangaea were teleported back down to the surface one at a time. "SHIP, Dawn will provide your general course, let's go back to that deserted planet."

"Dawn, SHIP is all yours, provide your course and monitor our progress, it's an eight week voyage."

"No problem, I have been making an interstellar computer map of where we have been to date. I can basically just use this touch screen to plot our course anywhere we have been to date. SHIP updates the map with any large rocks it detects during our passage. If you zoom in far enough you will see the rocks. SHIP has been adding data from as far as fifty light years from our various voyages. As you follow the same paths, the detail is very exacting, but as you get further out the details get vague. At one light year, it takes a very large rock to show up on the map. At fifty light years, it takes a planet the size of Mercury to show up at all. The stars that were discovered from all of the Martian and Earth history are shown. The map is three dimensional, but the touch screen is only two dimensional so you have to give a rolling motion to rotate your view."

"Why not have SHIP dim the lights in the room and project a holographic image which you can tell ship to zoom in or out. That way you can see a three dimensional known universe and use your pointer to draw the desired course for SHIP to follow?"

"Can SHIP do that?"

Karen interjected, "If SHIP can produce a good steak dinner out of raw atoms and make it appear on a china plate in front of you at the table, this holographic interactive display should be a snap. SHIP has been able to produce whatever we could dream up in a working model with improvements over what we dreamed up and do it instantaneously from orbit to any line of sight on a planet, this is nothing."

"SHIP?"

"Dawn, what room do you want to use? I can change the image from room to room or zoom it to cover all the empty space so you can walk through the known universe from room to room."

"SHIP, lets use the living area as it has more open space. Display only the part of the galaxy between Pangaea and Vulcan and Pyramid." The living area was a large somewhat round room with a sunken area surrounded by deep luxurious sofas built into the area facing a wall painting and fireplace which would transform to a huge high definition full screen television when desired. The curved outer wall appeared to be a glass wall with a view of actual space. It was not glass as it was part of the subatomic very dense structure of SHIP. If there were any dangerous rocks in the area that might hit the SHIP, the wall would instantly become a solid wall instead of glass. Larger objects that might be felt as an impact would be deflected using SHIP's anti-gravity. The "glass" would filter the light if close to a star that might be too bright and would filter less when not near a bright light source. The floor was covered in a very deep carpet like the bedrooms. Naturally, anything spilled on it would just disappear into the subatomic structure of the SHIP…no five second rule, it just never quite landed.

"Show Mercury size objects in that view so we can see just how full it is."

"Whoa! Too much. That was nearly solid. Display solar systems as points on the display showing our three colonies as green points."

"Thanks, that is more manageable. Now bring up the new planets as red dots. Now zoom to show our location and the nearest red dot. Okay, bring up more detail to show individual Mars size planets only between us and the nearest red dot. Okay, please draw a blue line at the end of my pointer as I walk our course."

Karen and Craig had just sat on the sofa watching this interaction between Dawn and SHIP. "Now isn't that a lot better than a two dimensional touch screen?"

"Yeah, Craig, it is. SHIP, do you detect any problems with this course?"

"As long as you will let me guide us around the larger rocks, there should be no problem, but remember, that even objects in space do move so the map will be changing with every voyage. I will keep you posted should new objects, worthy of note, be found."

"Okay, SHIP, all speed ahead."

"Okay, Craig, Karen, Dawn. I'll have you know that our Milky Way Galaxy is approximately one hundred thousand light years in diameter and thirty thousand light years thick at the middle. It is mainly empty nothing, but contains between two and four hundred billion, with a B, stars and probably ten billion, still with a b, planets in the habitable zone from its star or stars. We have been exploring only one of the outer arms of the Milky Way."

"SHIP, display a low detail image of the milky way that fills roughly fifty percent of this room. Thanks. Now show Earth as a red dot and our three colonies as green dots."

"Too small? Slowing zoom in on Earth like we are getting closer without increasing the display size. Craig, notice how the center of the galaxy is sliding away and disappearing as we move in on our outer arm of the galaxy. SHIP, hold it there. Karen, see how we are only showing the outer two arms of the galaxy yet we can't really see our location. Okay, SHIP; zoom in to where our portion of the arm of the Milky Way fills the room. Now brighten Earth and our three colonies. Okay, slowly zoom in where Earth is near the wall on my right and Pangaea is near the opposite wall. Show all known planets and moons in this view. Now brighten planets in the habitable areas and fade out the gas giants and all planets not in a habitable area. Okay, thanks, SHIP. Cancel display."

"Thank you, Craig. That was pretty neat and beats the heck out of a two dimensional display screen. Why didn't SHIP offer to do that while I was working on the system?"

"You have to remember that SHIP is not alive. It has stored knowledge and amazing powers, but it doesn't actually think for itself. It has programming from Atlantis supplemented by all humanity, but it cannot come up with anything new. Ask what you want of SHIP and SHIP will either produce or explain why it cannot. It can only produce smaller objects. It can produce a small airplane or even a small airplane size spaceship that can double as a submarine, but it cannot reproduce itself. Every time it uses energy its mass gets a little less dense. To recharge its mass, it must be near an object with mass and teleport some of that mass as subatomic particles to store as part of SHIP."

"SHIP, I have a question. I see no reason to, but could you take us to a galaxy far far away?"

"I do not know. My only knowledge comes from Mars, Earth and this voyage. At the distance between galaxies, I calculate that I would lose fifty percent of my mass to get to the next galaxy and would need a week near an object with lots of mass to recharge myself. That assumes that there are no objects in between galaxies. I could not skip over a galaxy but would have to make a stop there. However, even at my speed, it would take several hundred years to cross the known universe. While you would not age, you might get pretty bored."

"Okay, thanks SHIP. Dawn, Karen, I don't believe we will ever begin to explore the Milky Way. What do you think, Karen?"

"I go where you go, but I agree, after seeing Dawn's view of the Milky Way filling the living room, and how miniscule our little colonies are and how remote, NO, I cannot imagine ever leaving the Milky Way. Do you think there could be ships from other galaxies here in the Milky Way that maybe travel far faster than SHIP even?"

"Good question. Anything is possible. Less than ten years ago the capabilities of SHIP were beyond imagining, now, among the crew, it is routine. We have gotten used to SHIP's capability. The universe is forever in age, so there could be far more advanced ships out there. It has been tens of thousands of years ago that the Atlanteans built SHIP, so their technology may be far beyond SHIP's."

"Karen, Craig, it is getting a little deep in here. Let's get back to our current mission. I think our current mission is to visit more than one of the dead planets we found and try to find some evidence as to why there is no animal life around the cities, and explore more of those planets in case there is surviving animal life elsewhere on those planets. We should study one planet in detail, then move to another similar planet and see if they compare. Then we need to study the results to see if our colonies will face the same problem soon or far later, or never."

"Well, we are well underway now and have a couple of months enroute. Why don't you get your SHIP archaeology training out of the way?"

Dawn was in the training chair for four hours. Karen and Craig ate a hearty meal with some wine and then went to the master bedroom for some sport. They were waiting in the living room when Dawn came out of the training.

"Karen, your turn. That was intense. I am now specialized in techniques, pottery, bones, food remains and miscellaneous lithics in Archaeology. SHIP would not let me get intense into the written word. SHIP said it would teach Karen all about the written word and not let her get intense into lithics.

While Karen was in the training chair being trained in ancient languages of Mars, Atlantis, and Earth, Dawn ate while Craig watched.

"So, Craig, what do you think of having only two women?"

"The same as when we went to Vulcan last month. It was never my idea to have an all female crew even though I could not have done all this without them. How do you feel about it?"

"Actually, it has been somewhat of a relief. We married for convenience and I am not sorry for that. It gave my parents a good feeling about everything. It was just you and I for a long time so it followed that we would have sex regularly. The situation of all of us being cooped up in the ship for months at a time with only one man may have skewed all of our thinking. Yes, it was essential for the seven of us to form a team over several years to establish the colonies and save some few of mankind. Yes, being cooped up here for years together would lead to you being the sex object for all the supremely healthy females in the crew."

"You are saying that now that the colonies are established and some of the crew is off the ship for longer periods of time, the attachments may normalize and some of the women will drift away from me and towards other men?"

"That is what I am saying. All of us are still necessary to keep the colonies going and on track, but we no longer need to live on the ship and may find other romances as we rejoin the societies on the colonies. You chose each of us for our interests and rudimentary education, and looks, but you never had time to fall in love. Each of us women joined the crew for different reasons. You are not bad looking, and very smart, but probably not what each of us was planning on. None of us, including you, ever planned on creating colonies on new planets and exploring space and science."

“We created the colonies with a ratio of six women to each man in order to transport more people for a given weight on the ships and to provide for faster population growth in the new colonies. Yes, if we girls from SHIP were to live on planet, we would just be one of the thousands of women with a severe man shortage, and we would age again while we never age being teleported to and from SHIP. At the same time if we age another ten years we will probably not be as interested in sex. When was the last time Karen climbed a cliff? When was Shelby dancing with the other cheerleaders on a basketball court? SHIP keeps all of us as physically fit as our genetics allow and even our genetics has been corrected by SHIP to make us more than we were born to be. Sandy was a surfer girl. I think she may actually love you and be willing to give up being a surfer girl, but I would imagine that if given a chance she would like to live near the ocean again and have the fun of sailing her small sailboat in the surf or just hanging out with the surfers.”

“Dawn, as always, you are very logical and keep us grounded in reality. You are absolutely correct. Estera only came with me to escape Earth. I needed a botanist for supporting the colonies and onboard gardens on the ships. She was from a sophisticated political family. I would probably not appeal to her except for SHIP. What about you Dawn?”

“Dawn? Would you have been at all interested in me without SHIP?”

“Probably not, Craig. I was not interested in men period and have never had the desires of some girls. My experience before you was very bad and my experience with you has been beyond expectations. I am very comfortable with you.”

“You did not answer the question.”

“No, I guess not. Maybe if you had courted me over more time I would feel differently. Maybe if there were not five other girls involved, maybe I would feel differently. I’m sorry, I just don’t know if I am capable of true love or if the intense friendship with you is love. I really like you, I am comfortable with you."

"I think Shelby and Estera probably need to be with the general colony population more to see what it is like. Sandy needs more sun and fun. Jill is intense in everything. Right now her engineering skills trump everything else. I don't think she will even notice that there are other men around her. I would not be surprised if we lose Shelby and Estera soon. Sandy seems very much stuck on you, but would she be without the SHIP. However, you are not the typical surfer type or a body builder walking on a beach. Probably not. Karen was a loner before you came along. Jill. I don't know."

"Now I have depressed you. I think one of us will want to stay with you forever. I have no desire to go looking elsewhere; I just may not satisfy you as the only one. I think Karen will stay with you for the adventure. She may be more in love with you than even she knows. I want to stay with you, but I suspect Karen is more intent on keeping you handy. Jill will eventually run out of engineering challenges and will need to return to ship to rejuvenate so she will stay with you. Sandy gave up men in the years we were together before you decided to add her to the expanded crew. It may have been SHIP that got you together, she may want some sun and surf, but I think she really loves you. You might consider going on some sun and surf vacations. I think you are right that Shelby and Estera may very well find someone else. It was essential on Pyramid for us to be seven. It may be essential in the future. If we lose a couple you might have to replace them in case we come across another puzzle like on Pyramid. Right now be happy."

"I could not be more satisfied given the situation so far."

"So, how about you, Craig. How do you feel about me?"

"I found you first and quickly settled on you. I liked your looks, I liked you. I have to admit the decision was based on your looks and interest in interstellar navigation and your desire for space. SHIP had no experience or stored knowledge of navigation outside Mars and Earth. Oh, it had all the knowledge of Mars and Earth in astronomy, but that is not the same as navigating in space. Can you use a compass in space? No. Can you use a sextant? No, not when the star field is changing as you travel. The stars we see today will not be seen with the naked eye in minutes at the speed we are traveling. You were the one that made our navigation possible. I guess, with your star map of our travels, I could navigate through our known space. When we started, we just pointed SHIP in a general direction and traveled until we found potential planets, but now we really need your navigation skills."

"Craig, I asked how you felt about me."

"Like you said of me, I am comfortable with you. Under normal circumstances I would have been loyal to you forever. I was a monogamous kind of guy with no interest in more than one woman at a time. Am I really in love with you? I agree with you, I don't know. I am comfortable with you."

"How about the other girls?"

"Estera is not really my type. SHIP guided me to her for her father's position and her specialty in botany. It would probably have been a different botanist if not for her father's connection and SHIP's insistence. Maybe under normal circumstances. We do have similar likes and dislikes. She is attractive. I like being larger than she is, versus Karen is nearly my size and without the SHIP augmentation, she could easily overpower me. I would probably be afraid to approach her. Karen is one sexy Amazon. When I see her on alert for danger, she takes my breath away."

"Shelby had my attention as a cheerleader, but she is sort of withdrawn and cooler. She has a great body, beautiful face, and a gracefulness that is very compelling, but I would probably never have approached her in real life. I would have figured that she would never show an interest in me as I am not tall, not rich, not bad looking, but too ordinary."

"Sandy? Again, I would never have met her except SHIP sent me to her. She was in mortal danger, she was goddess sexy and I had no choice but to save her. I was so attracted to her physically, that I stayed away from her for years afterwards for fear of forgetting the mission. I guess she stayed away from everything hoping I would come back. I would never have seen her again except for your very wise idea of finding five other women to share the task with. I still held off on adding her. Love at first sight? Yes, I guess so. Would I have thought of her had it just remained you and I? No. I had written her off as too dangerous to the mission because of my instant attraction."

"Jill? I only met her because she was your best friend. She was attractive and any guy gets weak when flirted with like she did. Naturally, she attracted me physically and mentally it was hard to forget her flirting. I added her to the crew because it was easy, because she was already your friend, because she had the intelligence and an interest in science that I needed. She fit the bill after you. Love? No. Would I be happy if she were the only one? Yes. The physical attraction and like interests would have satisfied me."

"Let's get back to you, Dawn. You are my closest friend. You are my stabilization. You are my rock to fall back on. I need your intellect in more than just navigation. You complement my intellect. You are my sounding board. You are very beautiful and the one I originally selected. You are my only legal wife under Earth law. I do wish you lusted after my body a little more. I wish you would go for the sexy clothing. Do I love you? I'm not sure, but if the SHIP had never happened and if we were just students again, I would have tried to get to know you better with the idea of getting serious. Now, after years with all of you, I am confused."

"Okay, Craig. I appreciate that. I feel the same way. I do not want to look for anyone else. I am satisfied. Do I want you for myself without the other girls? Yes, I guess I do, but it sure would be boring with just you and me on the ship forever. We have spent most of our time for the past seven years on this ship. How many more years? Ten, one hundred years? This way, everything stays spicy. I would like to have children some day, but we can live forever. Maybe at some point we can settle on a planet for a few years and after the children are grown, go back to SHIP, and start all over, young again."

"I'm not sure SHIP would not select another captain, if I were to abandon SHIP for years. We have fulfilled SHIP's original mission to save some humans before the end of Earth so we are beyond its programming, but I don't think SHIP even knows what its next step would be. Our colonies won't be entirely secure for generations. The next generation will probably be an equal number of men and women so things will be much more normal, but it will take many generations to get the society operating in a normal fashion. There will likely be half brothers and sisters getting married and possibly upsetting the gene pool. Like the old jokes about rednecks marrying their cousins. That will be inevitable. We don't know that there won't be other threats to humanity. It would be nice to be normal on Earth with no specific fear and raise a family, but Earth is gone. Let's just keep going for now."

"Oh, I agree Craig. We have to keep going for now. If at some point we get tired of this in a century or so, we could always see if SHIP would find another captain."

"Dawn, don't take this the wrong way, but it has been years since I have been able to just sit around and talk with other guys about whatever. There are no politics, only panic and work. We have not discussed religion. There have been no football games, NASCAR races, or any sports at all. Maybe, when we get back we should organize some entertainment for the colonists if they have not started something on their own. Baseball maybe. I never liked baseball that much and never had the opportunity for soccer, but some sort of sports. On board the colony ships during the trip they had musical groups and dramatic plays, but we, the crew never took part in any of it. I think it would be good for all of us to take a little time off."

"Oh, I agree. If it weren't for the other girls I would scream at the walls. We talk together among ourselves, but I can see where you must be busting at the seams. Ah, there's Karen coming out of training now."

# CHAPTER THREE

# DIGGING UP HISTORY

"Karen, why don't you and Dawn discuss what you will do when we get to the planet? You two are the experts. I'll just listen and insert dumb comments or suggestions."

"Dawn, I will be looking for written language. There should be recorded media that did not disintegrate. We may find a library, a lab, a government agency, or a private residence that has some form of recorded media. It may be on a computer chip made of silicone and copper or some metal other than Impervium. Impervium is a total insulator, so it cannot be used in a computer other than the shell so there has to be other matter. Paper? A stone tablet?"

"We don't know how SHIP stores data. SHIP, if we find a similar computer to yourself, could you read the information?

"As far as I can know. I have never been exposed to another computer like myself. Also, keep in mind that my technology is tens of thousands of years old. That there was no advance in technology would not compute. Comparing my computer skills to the latest in the galaxy might be like comparing an Eniac computer with a modern super computer."

"Okay, I will be looking for written or recorded word and will just have to hope I or SHIP can recognize it for words. What will you be doing Dawn?"

"I will be looking for basically everything else. I will be looking for traces of their cookware, remains of not made of Impervium, traces of their food. I will be looking in the remains of the housing. I will have SHIP make me a super selective and sensitive metal detector that I can just carry with me, transmit information to SHIP and then have SHIP warn me if the detector finds something I might be interested in. Then I will dig for it or have SHIP expose it. If it seems to have writing on it I will use the communicator to tell you and have SHIP mark it in memory to make it easy to find again. Sort of like the galaxy map I had SHIP display. I got some training in recognizing written Martian, but you are the expert."

"Much of their ancient language was like the hieroglyphs in Egypt and elsewhere. We may find a history engraved on the actual stone building or decorating blocks in hieroglyphs. If they had a warning, they might have deliberately produced the warning in hieroglyphs to be a visible warning for thousands of years."

"Dawn, we will need some of those folding army shovels, trowels, brushes, and at some point we may need a front end loader to dig a deep hole if we find a basement or something that needs explored."

"May I interject?"

"Sure, Craig."

"Why not use your communicator to tell SHIP where you want to dig and then SHIP can teleport the diggings to replace its mass. It will need to replace mass anyway after the trip, so why not just sort of delete the hole down to the depth you want to inspect."

"Great idea, Craig. Why didn't SHIP offer?"

"Because SHIP does not think, it computes. That is a new idea for replacing its mass: directing the teleportation to do double duty of replacing its mass by taking it from where humans direct. In the past, SHIP has just flown over a desert and sucked up some sand in subatomic form to replace its mass. This is just having a human tell SHIP where to dig."

“Great idea. I understand. SHIP can do almost anything, but like we have discussed numerous times, we have to come up with the new idea.”

“Dawn, we will still need to be careful that SHIP does not teleport and convert some important evidence. We’ll just direct SHIP to suck up the mass until it gets close to where we want to dig and then use our manual tools for the final few inches. SHIP will be acting as hired diggers for shallow excavations and as a front end loader for the deeper digs.”

“Right, Karen. Good thinking. We will have to be careful what we tell SHIP to do. If Craig were the archaeologist, he could just think what he wants and SHIP would reproduce that exactly, but since we have to use the communicator to SHIP to describe what we want. We will need to go slow in most cases and be very specific in our wording.”

“SHIP, can you extract one inch at time if we mark off a grid?”

“Yes, Karen, but you need to stay at least twenty feet back while I do that. Actually I could be accurate within a small fraction of an inch, but I say twenty feet to make sure that you don’t accidently get in the way or fall in while I am extracting.”

# CHAPTER FOUR

# WHERE TO START

It took three months to arrive at the planet they decided to call "Destry", short for destroy. The two month course was blocked by a previously unknown solar system that had drifted into their proposed path.

"Well, we are approaching the planet. Where do want to start your archaeology?"

"Craig, let's orbit the planet a few times and let SHIP tell us what it might find rather than going directly to that destroyed city."

"Good idea, Dawn. I would have just jumped in where we left off without searching. I think you have the patience for archaeology more than I do. I have the training too, but I am more a doer than thinker. Too many years alone exploring the world on foot, I guess."

"Karen, you probably have the highest IQ of any of us. I have never quite understood why you dropped out of school and started your wandering."

"It's hard to understand myself now. That was several years ago. I dropped out of college because the professors were trying to brainwash me and I decided I would rather be an explorer and guide outdoors rather than an inside environmentalist liberal."

"Okay. Not wishy washy at all. Well, you are exploring now, but I guess you realize that you are losing your native complexion and starting to look like that indoors person."

"I am not. I am naturally darker skinned from my Apache ancestry."

“Look in a mirror. Sandy was darker than you are now when she first came to SHIP.”

Karen did go look in the bathroom mirror. “My God, I am turning into a paleface.”

Karen came back in and addressed Dawn, “Well, I will be outdoors on this archaeology expedition and get my tan back. I did see how pale Sandy was getting from being cooped up on the SHIP; I just never expected it to happen to me too. How come you, Shelby, Jill and Estera still look the same?”

“Makeup. Estera was always pale and an indoors person living in northern United States so she has remained pale. Jill and I were in a sorority in college and required to wear makeup daily. Shelby was a cheerleader who had to use plenty of makeup to match the other cheerleaders. It wouldn’t do to have one tanned and one pale cheerleader next to each other.”

“Well, I never thought I needed makeup. I don’t need to color my eyebrows or eyelashes as both are thick and black. My skin has always been ruddy because of my background and I guess time spent outdoors.

SHIP entered orbit and started mapping the surface looking for anomalies besides the one destroyed city they had already briefly visited.

On the first pass they identified three continents in the temperate zone and a small continent at each pole. The planet was one point four times Earth size but not as dense leaving gravity almost the same as Earth. They found fifteen cities on that first continent, all of which seemed to have suffered the same fate as the first one they found. There were hundreds of towns, all destroyed. The next continent was deserted, but most of the buildings still stood. On that continent there were still vehicles parked around the streets, but no evidence of people. They were getting infrared signatures of animals, but nothing human.

"That is very interesting. It appears their metal is still intact. I suggest we land there first to see if it is Impervium."

"Karen and you can go, Craig. I'll watch from SHIP. I think someone needs to stay with SHIP to watch the big picture. If we are all concentrating on a hole in the ground we may not see that we have something sneaking up or maybe we have weakened a wall nearby that could fall on the diggers. Once you have determined it is safe and ready to go digging, I'll come down and Craig can watch from SHIP."

"But, Dawn, if it is Impervium and has not decayed, maybe we don't have anything to have to worry about."

"I disagree. It adds more questions. Was there a war between the continents that resulted in the destruction of Impervium by maybe this continent? But where are the people. That also would mean that the people we are following are not necessarily friendly and they have weapons that can disintegrate Impervium. And, what of the people, were they just sucked up into their ships' structures just like our ship sucks up sand or dirt or whatever and incorporates it into the ships' subatomic structure? Gross."

"Karen, Dawn, let's complete the planet survey and get busy with the archaeological part of this expedition. Maybe that will answer what is not answerable right now from orbit."

"Okay, you and Karen go down to that intact city and inspect the structures and vehicles. I am staying here."

"Don't worry. We'll stay together the whole time we are down there and then come back. We will wear our communicators and keep them turned on so you can hear everything. We will also take some weapons, just in case. Okay."

"Karen, you go first with your weapons and I'll be there a few seconds later. You are faster at recognizing and taking care of any problems."

# CHAPTER FIVE

# THE INTACT CITY

"Well?"

"Haven't seen anything move, but with the vehicles and buildings totally intact everywhere, it is spooky."

"Yeah, I see what you mean. Still, most of the vehicles were over by that spaceport or whatever it was. I think that everyone was evacuated from this planet. From the looks of the other two continents, everything and probably everyone was destroyed."

"Let's just walk up the street toward what looked like city center and get an idea of this place. Craig, let's stay in the middle of the street, just in case."

"It is really deserted. Note how these vehicles are boxy. That gives more interior space within a smaller package. I want to inspect one a little closer."

"Okay, take that four passenger vehicle. I'll keep a watch out. Tell me what you see. Dawn, are you getting this on SHIP?"

"Yes, I can hear everything and see what each of you is looking directly at on the monitors."

"It was made for tall people. At only five ten, I would be small."

Craig could see no controls, just seating in the small four passenger vehicle. It had large vertical windows, obviously not built for aerodynamics or aesthetics. It ran on some sort of flexible tires, four of them, like rubber except not worn at all. Not rubber which would have worn. Maybe it was self repairing like SHIP. The motors were not visible, probably made of Impervium so they would never wear out. Lubrication was probably totally sealed inside the motor or transmission. Without wear, the lubrication would never need changing. Sealing it would prevent dirt and water and any contaminants from harming the oil.

"This vehicle was either guided remotely, programmed for a course, or used some sort of verbal or even thought control to guide it. I suspect it is like those "Google" cars they were experimenting with on Earth. I see no inputs for charging batteries or refueling it. Maybe it ran on electricity transmitted from wires under the roadway"

"Craig, maybe they used transmitted electricity like Tesla tried?"

"Could be, Karen. What do you think, Dawn?"

"We need to study more."

"Craig, SHIP has pointed out a number of animals roaming around the city and the countryside on this continent. It would appear that what you would expect without human interference. I have had SHIP zoom in on some of them. There is a full spectrum of mammals like deer, cattle, dogs or wolves, cats, rats, birds, insects. I cannot identify them, but they do not appear to be all that different from Earth life. There are trees of all types ranging from stunted aspen type trees in the colder areas and palm type trees in the warmer areas. If those other two continents with the destroyed cities were not so lifeless, I would see no abnormality except a civilization strangely devoid of people."

Karen said, "The animals and birds must not come into the city because there is nothing here for them to eat. Everything is stone or concrete and without people there is no food or it was eaten centuries ago. The city is large enough that animals would only come here to eat other animals or hide from predators. Look, up on the rooftops, I do see birds and what look like nests. Perfect for the birds to be safe from predators that might threaten birds and birds can fly and cover more distance quickly to go find food outside the city and fly back here to feed their young and nest in safety."

Craig asked, "What about insects? Wouldn't the birds find insects in the city to eat?"

SHIP replied, "The city is devoid of plant life and soil except where it has blown in. There are a few plants growing in side streets where dust has blown in and gotten thick enough for blown in seeds to grow. The civilization did not have plants in their city, everything is paved or buildings, but centuries of blowing dirt have completely covered some side streets and plants and insects that live with plants have gotten there. The main streets you are currently exploring are designed such that the natural winds keep them swept clean so plants cannot grow. If you explore side streets you will see plants and insects. Many of the birds you see eat the berries and insects of the plants. Some birds prey on other smaller birds. From space I cannot see any animals in the city, but that does not mean there aren't any."

"SHIP, show me an infrared examination of the area of the city near Karen and Craig."

"Dawn, you see the warm areas on the side streets near them. Apparently, there are some curious larger animals surreptitiously watching the strange humans."

"Karen, Craig, watch out, you are being watched but because of distance and buildings SHIP cannot identify what the animals are, but they are within two hundred meters of you watching you from the alleys and open doorways. They do not remember even seeing such creatures as humans, but if they are hungry they might decide to explore closer."

"Right. I have my weapons. Craig is going the exploring and I am watching him."

"And I will be watching from SHIP."

"I'm going into this building. Karen, keep watch by the door."

"No, let me go in first, we don't know what we will find."

Craig followed on her heels. “It looks like they had ordinary furniture, large by our standards. Apparently, they were from a taller race, maybe because of the lesser gravity of Mars before moving to Earth. Millions of years in evolution do not get erased in thousands of years of higher Earth gravity and I guess fifty thousand years on these other planets. Look here, Karen, these appear to me computer monitors or televisions they left behind.”

“Their furniture appears to be well preserved. Even the windows and doors are intact. It looks like we could just move right in.”

“Except for no power. I wonder if this planet has the natural resources like oil and coal. If so, this could be another planet we could colonize.”

“Not until we see the other dead continents. We don’t know if the same thing will happen here or if they just got bypassed by whatever hit the others. Maybe there was a war between continents and this continent was not involved, but what if it was a natural phenomenon that caused the devastation on the other continents. It could happen here at some point, or maybe not.”

We won’t know until we find out more. This must have been an elevator. Are there any stairs?”

“I don’t see any. Could they have been so confident of their construction that they had no need for emergency stairs?”

“Could be. Notice how the chairs and sofas appear to be perfectly preserved. Let’s see if we can find anything written to learn more.”

After a few hours of searching, they found nothing. Apparently, the whole ground floor was a meeting area with no offices or personal spaces. “We would have to get upstairs to find more. “SHIP, can you teleport both of us to the second floor?”

They were on the second floor. These were apparently offices and from the computer or television screens apparently paperless. Without electricity, there was nothing they could find out. They had SHIP teleport them to other floors but still found nothing in writing. The day was gone and they teleported back to SHIP to eat and rest.

"Tomorrow, let's explore some residential housing and see what we find."

"Yes, but I think it ought to be Karen and I while you stay in the SHIP. We were both trained in archaeology and may spot things you don't. Plus you can see on the monitors and hear whatever we see and hear."

Karen was tired and Dawn snuggled up to Craig in the master bedroom.

***

The next morning Karen and Dawn went down to explore residential housing. Craig watched the monitors to see what they were doing and kept an eye on the larger picture from SHIP's sensors.

"Karen, Dawn, I am seeing other infrared signatures near the house you are in."

"We'll keep the doors closed. See if you can figure out what is out there and let us know. So far we still haven't found anything written or powered. I am going to send up some of these tablet devices and a television or computer monitor up to SHIP for further study. We keep seeing them in the offices and residential rooms. Maybe ship can power them up and we can get some knowledge from them."

"Karen, I think this a bedroom for children"

"Obviously, but apparently the children are all different ages because ten beds range in size from grade school to high school size children."

"If four children is representative, they must not have believed in birth control."

“Back in our history they wanted lots of children to work the land. I think, they wanted lots of children to populate multiple worlds. When you have unlimited space travel and no other humanoids there is no limit to anything.”

“Right, Karen, the more planets you can colonize, the better the chance of your species surviving forever.”

“There is plenty of space in this house so apparently having separate rooms for separate age or sex kids was not an issue. Just lump them all together.”

“That was the way of our ancestors; the older children help train and care for the younger children.”

“Karen, SHIP has zoomed in on those animals outside the house you are in. They are either wild dogs or wolves. Can you see out a window and see if you can decide which and are you secure there?

“Dawn has decided that the windows are sealed with a glasslike material that is quite strong. I hit one with a chair and it just bounced off. The door latched solidly, so I think we are perfectly safe. I am looking at what is apparently a pack of wild dogs. They look well fed, but they are all larger dogs and have had no grooming for a few years and some have healed scars.”

“Karen, Craig. While Karen was looking out at the dogs, I explored the master bedroom. I think that the humans were evacuated from this continent in a hurry and they just left everything behind including their family pets. I have found closets with clothing still there. The master bedroom, with a bed for two has storage drawers filled with non-hanging clothes, underwear, bedding, etc. We have noticed how tall everything seems to be. If the clothing is any indication, the man was six-four and the woman six foot even.”

“Dawn, I’m curious about the kitchen. Have you seen any utensils?”

"No, I haven't. There is a table and chairs, but no evidence of food storage, plates, knives, forks, anything."

"I think that maybe they just thought up a snack and the house provided it, just like SHIP does for us."

"That makes sense. Maybe they did not read and write anything, they just thought it with their central computer teaching the children everything they needed to know as adults."

"Karen, Dawn, SHIP disagrees. That would make a society of automations with nothing new. Humans would simply die off from boredom without learning and doing. Humans need stimulation. With a central computer providing all the training and producing everything for them their musculature would atrophy and without producing thoughts of their own their brains would shut down."

"That would mean that those tablet devices must have been reading and input devices. Cooking was something they forgot how to do. We have seen fields and orchards that were apparently tended by humans at some point versus just machines."

"SHIP, teleport us to something you think might have been a government building."

"Karen, look. This wall is full of photos so we can see what their faces look like. They look like our cave men."

"This is a wall of their historical leaders. With no reference to any Earth age designator I cannot tell when. I don't even know if this planet used time referring to Mars standard, some galactic standard, or this planets revolutions around its sun. We tell years from time prior to Christ's birth, BC, and AD after his death. Both are relative terms because we had no written exact reference for either one. In fact many think his birth was missed by 4 years and 4 months, but at least we have a designator to measure time."

“This wall is a succession of planet commanders. I guess president or dictator or king are human terms. Maybe they were a militant race? This opposing wall is heroes of the race, not just this planet. This is the woman that discovered the impending interplanetary collision on Mars centuries before it happened. This gave the Martians time to start moving to Earth. I can actually read Martian since my training on SHIP.

***

Karen and Dawn spent several days going though buildings without finding much more about the culture, history or science. They saw no reference to schools leading them to believe everything was learned through their computer training like on SHIP.

SHIP then started a survey using radar to search below the surface of the city for anything that might offer more information of the history or science of the city. The computers and what appeared to be computer tablets like on Earth, but no way to charge their batteries if they had them. SHIP’s analysis could not determine anything wrong with the devices, just no way to power them.

Craig suggested, “I think they were powered by transmitted electrical energy like the pyramids of Atlantis. Could you transmit some electrical power through the air?”

“I have never done that, nor was I built to do that. I do not even possess that technology in my memory, but I do know that the Martians/Atlanteans used transmitted electric energy to run their entire society: airplanes, communications, heat for their houses, and what you would call automobiles. I have recordings of that which I could display if you like.”

“Yes, display it on the large monitor in the living area.”

The videos displayed the main city of Atlantis archipelago. Karen said, "I can understand most of the words from my training, but they are too fast for me to say them. I will provide a summary at the end."

They watched the fifteen minute video which pretty well showed them what was happening and then at the end Karen paraphrased, "They never said how they generated the electricity or transmitted it, but the pyramids were topped with an omnidirectional antenna that beamed electrical energy from down the sides of the pyramid to all directs. Aircraft and vehicles were powered by direct energy in line of sight, but as they got further away from the antenna, they were powered by energy reflected from the ionosphere like shortwave radio used by Earth's amateur radios. The power from the antenna was directed in varying strengths depending on the receiver. The maximum distance was about twenty-five hundred miles so their aircraft would have full power line of sight, but gradually slowing as they went over the horizon relying on reflected power from the ionosphere which was much weaker even though most of the power was directed upward to reflect the power. Toward the limit of twenty-five hundred miles another power pyramid was required to provide power. Consequently ground vehicles were limited to a few hundred miles whereas aircraft could go well beyond that. Electrical power for their computers was sufficient for the entire twenty-five hundred miles, however, the energy for heating was similar to that of ground vehicles and required much larger antennas. Apparently for distance the power from the pyramid actually heated and concentrated the ionosphere like our military HARP system did in Alaska. By transmitting large amounts of energy at the ionosphere they could reliably bounce the electrical energy to the full distance."

When Karen finished Dawn suggested, "Then we could determine that Atlantis must have been within twenty-five hundred miles of the pyramids of Egypt?"

Craig said, “Not necessarily. The pyramids of Egypt were thought to have been built long after Atlantis disappeared, but on the other hand, there are pyramids all over the world that might have had generating and transmitting antennas in ancient history. There are potential pyramids in Central America and potentially in places as far away as China. It has been thousands of years since Atlantis disappeared so the pyramids probably don’t represent much left of the technology they might have had back then.”

“Craig, Karen and I have several of the small pads. Why don’t we have SHIP whip up a Tesla coil and see if we can get one of them to react?”

“Excellent idea.” SHIP materialized one on a table and Craig used some insulated tongs to place a pad near it. When the pad got within six inches, the screen lit up with a photo of a family, so they knew they had the general concept, but no way of knowing what to do next. Was it thought controlled or a touchpad? Did it do anything besides display a photo? As Craig just held it there, the image began to move to show a home video of the family with the family dog, probably an ancestor of the wild dog pack Karen and Dawn had encountered.

Dawn reached out with a finger and touched the screen. Some sort of icons appeared on the screen. “Karen, any ideas what to do next?”

“Not really. SHIP taught me their language, but these are just stylized pictures of something. Apparently, this is similar to an Android tablet from our planet, but probably far more powerful. Maybe we can have SHIP place the tablet on some sort of bracket to keep it this distance from the Tesla coil and we could experiment more with it and see if we can make sense of it.”

Before they could do anything the screen went dark and they could get nothing more out of it. Craig said, “I think we burnt it out with too much power. I think tomorrow, you two should go back and collect forty or fifty of these things for us to experiment with. Maybe we can find an optimal distance to provide power without burning it out."

Dawn said, "Maybe that is what happened on this planet. Maybe their generators went wild and burnt everything on two continents out."

"Did either of you see anything that resembled receiving antennas on any of the buildings or something in the buildings that could be a power regulating device."

"No, but we weren't looking. Dawn, you know a lot more science than I do. That will be your job tomorrow."

Dawn and Karen reported, "In inspecting the pyramid, it looks just like the one on Pyramid including what appears to be the combination that requires six women and one man to operate. We don't have a full crew so there is no point in trying to get into it. All we could do would be to bring the whole crew back and try it. It would take two months to go back to Pangaea and collect the whole crew and then we don't know if any six women and one man would work or would it be coded for a particular set of authorized people. I suggest we continue our research here."

***

They spent a nearly two weeks wandering around the city while SHIP was searching below ground for signs of human construction. They collected fifty of the tablets and a half dozen of the televisions or larger computer screens or whatever they were. SHIP did not have that much storage space and since SHIP did not know how to make one they could not just be assimilated into its mass and reproduced like so many other things it knew how to do. Therefore, they had to stack them in a box that SHIP provided them and left near the window in the living area. While the women were searching the city, Craig was watching the underground images. SHIP could not image in or under the pyramid, and found nothing built underground. SHIP never left line of sight with the buildings the women were in, just in case they had to be evacuated immediately, but Craig was able to pretty well search the surface for 200 miles around the pyramid which took in a lot of countryside. There were a number of smaller villages near the city getting further apart as he looked further out. There were trees planted in neat lines with a lot of wild growth trees growing between the neat rows. There was little fruit and since the orchard had not been tended in centuries most of the trees were dead with many just leaving a mound were roots once were. He had seen deer and cattle and dogs and small cats, and the inevitable rats. The Martians had probably killed off any natural animals leaving only the wild dogs as predators indicating that the people had evacuated the planet in a hurry leaving their pets behind. There were hundreds of vehicles parked around what might be the remains of a large spaceport, which added to their conclusion that there was a mass evacuation of the planet. They had found nothing like a threat to the planet. It had not been hit by any large asteroid since humans had been there and certainly not since the evacuation. From the looks of the remains of fields and orchards and the animals, on the intact continent, it had been centuries since the evacuation.

# CHAPTER SIX

# DEAD CITIES AND DYING CONTINENT

After spending nearly a month on the deserted city they moved to the largest dead city on the nearest dying continent. Like before, they spent several days observing the continent looking at the layout of the cities, towns, and villages of the continent. Then they had SHIP use radar to look under the city center. They found underground rooms under a large square foundation that was several city blocks square.

Craig observed, "That looks like it maybe it is the foundation of a pyramid. It is right in the city center like on our planet we have named Pyramid. It is even larger than any other structure or remains we have found anywhere."

Dawn asked Karen, "What do you think? If that was a pyramid, maybe there is something underground where it once stood."

"I think so, but is it stable? To have multiple basements there must have been a lot of Impervium used as structural supports keeping everything including a giant pyramid supported."

"Karen, this is the only subterranean structure we have seen anywhere other than the mines left for us to find. We could not see under the pyramid on planet Pyramid or under the pyramid on the last continent. There is something that stops SHIPS sensors from penetrating. This is our chance to maybe understand some of the technology involved."

"But will we understand it without the rest of our crew? Craig, what do you think?"

"I agree with Dawn. This is a prime opportunity. Dawn is no slouch at science. I have a good general knowledge I gained from SHIP, and we have SHIP's knowledge. I say we go for it. Karen, I should go alone first. SHIP has shown that it can zap me back here out of a threat instantly, much quicker than it can you or Dawn. Dawn, you and Karen monitor what you can through SHIP. SHIP, zap me down to the first level basement."

***

"Okay, everything seems untouched and safe. There are a lot of monitors and controls here. It looks like this is the major control room for whatever the pyramid was supposed to do. Let me do some more exploring. I think I see stairs beside the elevator. Maybe they had backup stairs here because this is the power center so they had a backup."

"I'm taking the stairs. There are pipes and big control panels and valves. Do you suppose this is a geothermal generator station?"

Dawn replied, "Makes sense. There were other clues including from Plato who said their volcano blew up. Maybe it was an asteroid that destroyed Atlantis, but maybe Plato really did know a little about the real Atlantis. Geothermal power could be really powerful if harnessed. I'm sure they had the capability to drill to the planets core and then pipe water down through Impervium pipes, superheat it, and then run turbines to produce unlimited electricity without any pollution or using any carbon fuels or even nuclear. Whereas solar only works in daylight, assuming no clouds, and wind power only works when there is a wind, geothermal would provide free energy for as long as the planets core temperature stayed hot. Millions of years of free energy. The Impervium would never rust and is nearly indestructible and could withstand the extreme high pressures. The turbines would last forever."

"Okay. Everything seems perfectly safe here. Dawn, why don't you join me in exploring this inner sanctum."

The two of them explored the various levels and found what Dawn said "Everything is what I would expect of massive turbine generators and water cooling lines to pump back to the planet's core. This is far beyond what we could have done with ordinary metals we had on Earth. The pressure must have been absolutely tremendous. The generators, while massive, probably operated at ridiculous revolutions per minute. I don't know what the wires are made of. If they were copper they could not carry the kind of power to needed to power a continent without melting."

"We don't have the technology to cut any samples of any of this, but the piping looks like Impervium. I want to go back to the other continent and look at the antenna at the top of its intact pyramid."

"Not until we get Karen down here to try to read some of the controls. We cannot power up the monitors, but maybe she can make sense of the buttons and switches."

Craig went back to SHIP to be replaced by Karen in the first basement with Dawn to study what they assumed were controls. Karen stated, "These were the controls where they could regulate the water flow from the core to the generators to regulate the power. There is no way to know just how deep they drilled toward the core. Theoretically they only drilled down a few miles. If the core cooled they could shut down the system, drill down a little deeper, and have power for another few centuries. Even on Earth we drilled down to thirty-thousand feet with the center of the core at over two million feet. Not knowing the core temperature of Earth, let alone this planet, there is no way to know how deep they drilled. The drawback to geothermal wells drilled really deep is that there could be phenomenal pressure in the pipes. If it blew out there could be a big explosion … enough to blow up Atlantis. If they drilled down to magma, they could have a new volcano spewing magma from the center of the planet, but I don't think that is what happened on these planets."

***

They spent the next several days on SHIP surveying the continent, then the other destroyed continent and its destroyed cities. The big question still remained, “What happened to the Impervium that was presumed to be indestructible?”

“I think it is time to go see the other planet that we found that seems to have been destroyed and see if we find the same things.”

“Craig, I suggest that we go see Vulcan before going to the other planet. It is closer and I am concerned for the women there. We owe it to Li Wang for being honest with us. We exposed her to danger in telling their dictator how we wanted the women treated.”

“I agree, Dawn. Vulcan is actually closer than that other destroyed planet.”

“Are both of you in agreement? It will add over two months to our trip and we have already been gone for over three months.”

“YES.”

“Okay, SHIP, take us to Vulcan.” They destroyed another ten tablet computers trying to get the voltage right from the Tesla Coil. They had SHIP mount it on an arm that could be adjusted by fractions of an inch and they tried varying the Tesla Coil voltage, but failed. It took four weeks getting to Vulcan and they decided to take a break from their research and watch some earlier recordings of their trip. They started with their earliest trip together to Pyramid.

Craig was the first to comment, “Dawn, you look like you have aged several years. You are more beautiful now than when I met you. Karen, you don’t look much different than then, but paler.”

"You are right, Dawn has aged. She looks to be twenty-five now and maybe twenty then. Craig, you look like you have aged even more. I would guess you were twenty-one then and mid twenties now. You may not see it, but I have aged too."

Craig asked SHIP, "When I first came to you I was only nineteen, and now I look twenty-five. I thought you said we would not age?"

"You have all aged to an appropriate age. You and Dawn were only teenagers when you first came to me. Karen was only twenty. I allowed you to age to be fully adult and at your prime. It has been ten years since Craig was selected and brought here. His actual physical age is twenty-nine, but I have kept him at twenty-seven. Dawn was closer to twenty when she came here, but now I am maintaining her at twenty-five, which is her prime age. Karen, you have been allowed to age to twenty-nine, your prime age. From this point on, you will return to your current age every time you are teleported. Craig may have thought his body was mature at nineteen, but twenty-seven is his prime genetic age when his bones and musculature are prime. Karen, I have let you age a little longer because your outdoor physically fit lifestyle has kept you from reaching your prime until a little older than Craig. Dawn, I stopped your age at twenty-five because without modification, you would be showing your age at thirty-four. Your prime age is twenty-five which you will now stay at as long as you are teleported to and from here. If any of you were to leave for ten years, you would age ten years, but as soon as you teleported back here you would return to your now current physical prime age."

"SHIP is right. We look awfully young in these earlier recordings. It is amazing we got anything done."

"We would have been laughed at by adults if it had not been for SHIP's training and help demonstrating things to the adults of Earth. Karen, Dawn, if I were to meet you now as teenagers I would consider you too young. I would be much more attracted to you as you look now."

"Well thanks, I guess."

"Dawn, admit it. If you were at your age now you would agree with me that considering Craig now, you would think him a brash teenager not worthy of your consideration as a date, let alone a mate."

"Karen, Craig, wait until we show these videos to the girls we left back on Pangaea."

# CHAPTER SEVEN

# RETURN TO VULCAN

"Let's just orbit and observe for a day."

"Okay, but I don't know what we will observe. I am anxious to go visit Li Wang, assuming she is still in power with the women and has not been executed for being honest with us on our last visit."

Karen said, "Look at that spaceship. Large sections are missing out of it. Is the same cause as on the planet we left two weeks ago?"

"Good question. What do you think, Dawn?"

"Why just the one spaceship and why just pieces of it? If the same thing was happening here, wouldn't it affect more than just parts of one ship?"

"SHIP, look for Impervium somewhere it wasn't when we were here last."

"Craig, some of their lighter armored vehicles now seem to have Impervium plating over their regular armor plate. I also detect Impervium inside some of the wooden structures that appear to be warehouses or factories. I also detect very high concentrations of static electricity in those buildings."

"Take me down to one of those buildings in a spy mode where I do fully materialize. I do not want to be seen, but I want to see what they are doing. Be ready to zap me back if there is any direct danger."

"Craig, I am the Indian Warrior here. Why don't I go in your place?"

“Sorry Karen, SHIP cannot do the same with you. I can go invisible, not fully materializing. I can see and mentally connect direct with SHIP versus having to use a communicator. I have done it before.”

“Okay, but Karen and I will be watching whatever SHIP can show us.”

***

When Craig returned to SHIP he gave them a more detailed report. “They have found a way to cut the Impervium from the spaceship and are using it to build small, what I think are, spaceships. They have found some way to extract radioactive material to build small nuclear propulsion devices like on the big spaceships shielded by the Impervium. They are small ships for only two crew with room for food and supplies for a long voyage. They have found a way to bend and weld the Impervium into new shapes. As they have not touched their other spaceships and only part of the one Karen spotted, they are just starting with whatever they are doing.”

“I think it is time for Dawn and I to go visit Li Wang in the women’s compound and see if she will provide us some intelligence as to what they are doing. Karen, we should be okay there, but what they are doing is beyond what SHIP taught us with Impervium so I don’t rule out anything. I am concerned that they have become enemies of the rest of the colonies. I don’t know that, but if you lose contact with Dawn and I for twenty-four hours, tell SHIP to destroy everything human on this planet.”

“Are you sure? That is pretty harsh. In any group of a few thousand people, all but a few hundred are probably innocent.”

"Karen, we have had trouble with the Chinese from the start. Originally they took military hardware on their ships instead of what they were supposed to. Their intent was to overpower all the rest in the colonies. They did not know that we were sending them to a different planet with no knowledge of where the rest of the colonists were and no way of getting there. I considered just leaving them on Earth or destroying their ships enroute, and decided to give them this planet to keep them isolated. We might have to reconsider that and go ahead and destroy this colony for the good of the other colonies."

***

Craig teleported to the women's compound first in case SHIP needed to zap him back quickly. He appeared near Li Wang, the apparent leader of the women when they visited last. "Hello, Li Wang. We came back to check on you. How are you being treated now?"

"Craig. Is Dawn with you? I want to thank both of you for what you have done for us. We have everything you asked for us and more. Some of us are even allowed to work in scientific endeavors. When out of the compound, we have protection details, but total freedom to go where we want and do what we want. Most of the women have been content to just have babies to help the colony populate this world, but a few of us need more than procreation. Some of the women are now valued scientists that are responsible for some major breakthroughs."

"If it is okay with you, I will bring Dawn here to talk with your women."

"Absolutely, she is a hero to the women here. We have unlimited reading and video privileges, total freedom to go around the planet, plenty of food, crafts, and the few that are actually happy to be involved in higher learning and science. Keep in mind that most of the women were really chosen as female slaves, but they had to speak English and have some learning. Most have no aspirations to be anything but vessels for increasing our population, but even their plight is very much better than before. We see Dawn as the person that made it happen. There are few unhappy women here now."

Dawn appeared beside Craig, "Hello Li Wang. Are you still in charge of the women here?"

"I guess, but things are so different here that most of the women no longer need a leader. Everything you asked for us women was granted and most of us are as happy as we could be. We have it far better than the men in the colony. They are still regimented and controlled by the central colony government, but the women are free to do as we wish as long as we keep the birth rate growing."

"I see you have found a way to remove Impervium from one of your ships and have found a way to form it into new shapes and even weld it."

"Yes. Hua Ming was a Chinese international gymnast when the crew was formed. She was smaller than the typical woman in the colony, but extremely fit and intelligent. She has developed a device to cut, soften, and weld Impervium. We don't know how to make Impervium, but by dismantling our grounded spaceships we have a goodly supply of the raw Impervium."

"Do you know how it works?

"No, it was developed by Hua Ming. She is one of the few women unhappy here. We have had a hard time keeping her free from birthing babies for the colony. Her brain is what has helped me protect her. Most of our scientific advances were as the result of her experiments. As a gymnast, they gave her more education than the typical Chinese female. That was so she would represent the Chinese to the rest of the world and make it appear that all

Chinese women were treated with respect. She is very intelligent and analytic in everything she does."

"Why is she unhappy, you ask? She feels like she is like the scientist that developed the nuclear bomb. She is afraid her advances will be used for other than peaceful purposes. She does not want to be forced to have sexual relations with the Chinese men that come here to sire children. It is getting harder to keep her from what the government considers her duty to procreate."

Dawn asked, "Has she told you any specifics of why she is concerned?"

"No. Just the general comment that she is concerned."

"Would she talk with me?"

"Probably not, but she knows about Craig and since she was raised as subservient to men, she might tell him in private."

"Okay, Craig. Plan on spending some time alone with her. And Craig. I think we will be losing Estera as a crew member. Maybe you should consider adding her to our crew."

"Li Wang. How would you feel about her leaving to never return here?"

"That would be perfect, Craig, but you would have to tell the commander about it before you leave Vulcan to keep me out of trouble. She cannot just disappear. She is closely guarded except here in the women's compound and I think the commander has been looking to take her as one of his wives. He would have done it except for her general attitude about men and commitment to science. If she ever fails to produce in the science area, he will take her and there is nothing I could do to stop it. She would rather be dead. She would rather just be one of the baby producers than be taken by him. She really is unhappy here. Can you?"

"When she returns I will talk with her alone wherever you suggest and we will see what she thinks. Maybe she will tell me why she is worried and I can take care of it until next time we come back here to visit. When do you expect her back?"

"She should be back by dark. That would be eight P.M. our time, I have no idea what time you operate on."

"We don't really operate by the clock. When near a planet, we pretty much operate during daylight hours and rest whenever we can. On our ship, lights are regulated to simulate Earth daylight. Which means at least ten hours a day we have indirect lighting for whatever we are doing or no light at all except for the stars?"

"It is much warmer here than in China. We don't have winter and it stays constantly between seventy and one hundred degrees, perfect for dinosaurs. We have not encountered any large dinosaurs and we can handle the small ones, they taste like chicken."

"I think the commander is very happy that the women have freedom now. The men were not doing well at tending the gardens and trees we are growing."

"Some of the women are specialized to raising the test tube animals. In another year or two, we should be self sufficient without eating dinosaur at all."

"You will have to ask Hua Ming why one of the spaceships reactor is no longer providing electricity. I do know that we built some generators that are powered by diesel engines we removed from a couple of our tanks. It is only the outlying areas that have to use diesel for generating electricity. We still have our reactors for this main city and one outlying city."

Dawn came back in, "I am so glad we were able to help you. The women act like I am some kind of god to them. They all seem so happy, although I really do not understand their desire to have many children. Some already have two children and another in the oven."

Hua Ming came home to the women's compound early at 5:00 P.M. "I heard that Craig and Dawn were here."

Li Wang introduced them, "Hua Ming, this is Craig and Dawn from the master ship. They have come to check on how the women were being treated. I have told them about your interest in science but that you are not happy here. I could not really explain why you felt that way. I suggest that we invite them to eat with us in the dining room and then, if you want to you and Craig can retreat to a private room to explain your unhappiness. He might be able to help you."

Dinner was noisy as all the women wanted to crowd around Craig and Dawn to thank them again for their new freedom. As one said, "We were never treated so well back in China. Young girls are frequently sold off or married off without their consent. This is not true of the upper class Chinese women, but then there are only a few upper class women."

The dinner was good, but Craig's taste ran to sweet and sourer Chicken or dinosaur in this case. He was never big on the cabbage and sprouts frequently served. Dawn ate it all, "Do you have cabbages that you can serve them regularly?"

Li Wang replied, "Actually, this is wild cabbage that grows in the jungle. It does not taste like cabbage until we have prepared it, but it does make a reasonable substitute. The commander keeps much of the Earth type produce because we do not have enough yet to feed the whole population, but we may have by next year."

***

After dinner, Li Wang and Hua Ming whispered between themselves and then Hua Ming offered a hand to Craig and gently tugged on him to follow. Li Wang told Dawn, "Do you want to stay here or return to your ship and wait for Craig. I suspect he will not be returning until morning. Hua Ming is not a virgin, but close to it. However, she speaks perfect unaccented English from her time on the international gymnastics circuit. She was encouraged to mingle with the English speaking gymnasts and news media to promote the new Chinese culture. I suspect she will talk his ears off and then take him to bed. Oh, I hope that doesn't alarm you. I forget that Americans and Europeans believe in monogamy."

"No I am not alarmed. The standard for all the colonies was six women to each man in order to populate the new colonies faster to ensure their survival on empty planets. Even our ship has the same mix although Craig only has two women with him on this trip. The other four chose to stay behind on their own scientific missions."

"Why have we not met the others?"

"We usually leave someone on the ship to provide an overview of the situation and watch for trouble the ones on planet might not see until too close. Karen is half American Indian. Are you familiar with the term Amazon Women?"

"I have read about them in fiction books since we got full computer access. Is she an Amazon?"

"No, she is part American Indian, but tall and muscular and can be intimidating to those that don't know her. I am the most easy going and best to mingle with new cultures. I am more empathetic to others and am almost co-captain of our ship. Karen acts like our warrior protector when we are in dangerous situations. Have you seen any videos of a barbarian queen?"

"Yes, but they were very attractive."

"Oh Karen is attractive all right, like a well tanned goddess. Well, I am glad that everything is going well; I'm going back to SHIP. We will meet with the guy you call commander tomorrow to make sure things stay good for you."

"Thank you. Will we see you again?"

"I'm sure of it. We might even visit again before we are on our way."

***

Hua Ming led Craig into a private bedroom. "This room belongs to Li Wang. She said we could use it as long as we want, she will be sleeping in the common rooms tonight."

"I am here to find out what is troubling you about living and working here. Li Wang said you were quite the scientist, but were troubled."

"Yes, I am troubled. Yes, I am a scientist. Can we talk here without being heard?"

"Yes, my ship has set up a field around this room to make sure no one and no thing can leave or be transmitted from this room."

"I have made scientific breakthroughs that I am afraid are being used for the commander to build interstellar war ships."

"I saw some small nuclear powered spaceships being built using Impervium. I guess that means you have found a way to manipulate Impervium as if were mild steel."

"How did you see the spaceships? Never mind. I have invented a very high powered electron gun that interferes with the electrical bonds of the Impervium allowing me to use it like a cutting torch to cut out sheets of Impervium from one of our colony ships. I may then place it a very powerful electrical field that makes it semi-flexible where men can hammer it into new shapes. I can then use my electron gun to weld the new pieces together to create new items. It started as a way to arm our light weight armored carriers against the larger dinosaurs. Then I was asked to build new small reactors and come up with a way to extract uranium from the large reactor on the colony ship to provide fuel for these smaller reactors. These small reactors were to be used for providing electricity to far flung remote expeditions."

"Do you want me to call you Hua Ming or something else or some better pronunciation?"

"You may call me Jane. That is what I went by among the international gymnasts that had a hard time with Chinese."

"Okay, Jane, it sounds like your research was done for good causes. You have done things we did not think could be done, like cutting, bending and welding Impervium. How do you do it?"

"It is hard to explain. Basically I use an electron beam like sweeping the screen of an old fashioned radar display. Before flat screen televisions, all televisions used that technology. Now imagine a ground based radar putting out a beam of electrons less than a tenth of an inch in circumference. That excites the nuclear bonding in the Impervium and symbolically melts it at the point of the beam. The workers bending the Impervium need to have special uniforms to protect them from working in a very high power wider beam. When bending a sheet of Impervium we have to take the city off the nuclear power plant to provide power to the electrical field where we bend the Impervium. Again, we use the tight beam at a lower power to weld the Impervium to itself. That is grossly simplified."

"So what is making me unhappy? They are asking me to modify the beam for long range application to use against other ships make of Impervium. They want to use a new form of radar to track your ship to the other colonies without you knowing they are tracking you and send out the small spaceships with high powered beams to destroy your reactors and make the colonies easy marks for the Chinese power. They do not have a completed ship yet, but it is close and I have not built a high powered beam. I do not want war with the colonies. I do not want to harm you and your crew. If I do not do what they say, I will be made one of the commander's wives and lose my freedom."

While she talked Craig assessed her looks. Like most Chinese gymnasts, she was short, but had grown far too busty to be a gymnast. Naturally her hair was black. It fell just below the top of her shoulders. Her eyes also appeared black and just a hint of the oriental slant, but almost a round eye like a European. Her waist was very small with her hips almost as wide as her shoulders. She was muscular like you would expect from a gymnast. He had never considered having a Chinese wife, but he needed her talent and she looked just fine.

"Would you consider leaving the Chinese society and coming with us as part of my personal crew for the next hundred years or more years?"

"Oh yes! But I don't think I will live another hundred years."

"Normally I have weeks or even months, in a couple of cases years, but I need to explain what joining my crew means. With you and the critical time for you, I must be blunt."

Craig ran through the basic formula of each man having six wives or concubines and how his personal crew was the same.

"I know that all the colony ships were to have that mix of women to men. I think I like you. Here, most of the women have no choice but to sleep with whatever man is sent to you each night until well into a pregnancy. Then the men are slowed down considerably until at least a month after the birth of a child and then it starts again. I have never had to do that, but I would rather do that than be taken by the commander for one of his sex slaves. You are an attractive young man. I may only be four foot eleven and ninety pounds, but I was a gymnast and very physically fit. I can hold my own against your other wives. I have not had much actual sexual experience, but we were all trained before starting this mission on how to please a man."

"I agree that you are very attractive and, yes, you would be one of my wives, but I am more interested in your science. We have many issues to face and you have solved a couple already without us."

"We have Chinese and other oriental people as part of our other colonies, but you would be in a minority, but not totally alone in our world. You would have to commit sexually to be taken to my ship and commit personally for the rest of our lives, may hundreds of years."

"You speak as if hundreds of years are possible, and I do not believe that, but, yes, I will have sex with you. I would enjoy the sex more than any sex I could have on this planet. At least you are giving me the option, I would like to come and visit some of the other women here, especially Li Wang."

"It might be unwise for you to return here on the planet and Li Wang cannot come to our ship, but you can be with us as we visit in the future and see and hear what we see and hear as we visit. You will be able to speak with her by radio."

Hua Ming, Jane, took off her clothes for Craig and then came to him and started pulling off his clothing. She pushed him back on the bed and moments later they were on SHIP.

# CHAPTER EIGHT

# BACK TO EARTH AND MARS

Hua Ming, Jane, woke in the master bedroom on SHIP. She thought of wearing conservative clothing to meet the rest of the crew and her idea of clothing appeared on the bed in her exact size. She found the master bath, bathed, dressed in the clothing that appeared for her, fixed her hair using a new copy of her favorite hairbrush, which she had not seen in years, then came out through the dining room to the living area.

Dawn, Karen, and Craig were there watching a large video screen. Through the curving wall sized window she could see Vulcan from orbit. She just stood there amazed.

As always, Karen saw everything first and saw Hua Ming standing in the doorway like a small child amazed at what she was seeing, "Welcome Jane. Craig has already told us to just call you Jane. I remember when I was in your position and the uncertainty of deciding whether this was a dream or a nightmare or reality."

Dawn was already going to Jane and had an arm around her guiding her to the wraparound sunken sofa. "Yes, Who Ming, welcome to our little family. I am Dawn, that is Karen, and you really know Craig as a wife should or you would not be here."

"Actually H-U-A is pronounced more like how woo ah, but Jane, just call me Jane. I was concerned about being a late added wife. Where are the other wives? I thought there would be six of you?"

Craig answered, "This is a research mission. The rest of my crew stayed on Pangaea to work in their own specialties. This was supposed to be just an archeological mission to study a couple of planets deserted by our predecessors centuries ago. Dawn insisted we come here to visit Vulcan and see if the women were being treated right."

"What is Pangaea?"

"Sorry, I should explain. We put earthmen on three different planets to help mankind survive. There is Pyramid, Pangaea, and Vulcan. Pyramid was named because there was a large pyramid in the middle of an empty stone city. Pangaea is named because it has one large continent like the mythical Pangaea continent of Earth. Before I go into lengthy explanations, you need to visit the learning chair where the history of Earth and Mars and what happened to Earth will all be explained. We have months before we get back to Pangaea so you will understand a lot by then. But we will wait until tomorrow for your learning session. Right now I'm hungry so let's all eat and be on our way. SHIP, take us directly to Pangaea."

Dawn led Hua into the dining room with Karen and Craig following. "Hua, uh Jane, just think what you would like to eat and you will have it."

"I don't think you will know how to prepare it and it would take hours to cook."

"Trust me, just imagine it." She did and SHIP made it appear in front of her cooked just the way she imagined it.

"What just happened?" Then meals appeared in front of Dawn, Karen, and Craig as if by magic. "I saw that happen, but is this a hologram?"

"No, Craig is the master of this ship which we named and address as SHIP. SHIP is intelligent. SHIP reads our thoughts and created this food from subatomic particles to exactly replicate what you were thinking. I believe that you will find the food just as you imagined. Each of us received something different because that is what we imagined and SHIP placed it here on fine china. The same with your drink. SHIP knows everything in the history of mankind. Every book, movie, recipe that mankind has documented. You will learn more of this in the learning chair tomorrow. SHIP will give you the history and training in your field. What is your field? Craig can communicate anything mentally with SHIP, but SHIP can only read our minds here on SHIP and then only what our likes and needs are. Otherwise we have to talk to SHIP like a person."

"I was an international gymnast representing China in the Olympics years back. My training was in electrical engineering. It took 3 years from what you would call high school to get my PhD. This was nowhere near as advanced as in an American school, I know, but I was chosen because of my fame as a gold medalist gymnast for China and because of my proclivity in electronics."

Craig interrupted, "I chose Hua, Jane, because she has invented an electronic beam that can interrupt the electrical bond of Impervium allowing it to be cut like with a cutting torch and welded into new shapes. It allows taking the space ships and other things manufactured on earth and making them into new things. It could also be used as a weapon against things made with Impervium if further developed. It is a technology we can use that goes beyond what SHIP knew and by having her with us, keeps the technology from Vulcan. She has also developed small nuclear reactors and other things that we never even considered."

"I was resisting that further development on Vulcan, but why is it important to you to keep Vulcan from having that when you want it yourself?"

"You will learn the history of why the Chinese were sent to Vulcan in the learning chair, but to put it simply, we isolated the Chinese on Vulcan because they did not follow the direction we gave to all the other countries. We told them to send six women for every man. They didn't. We told them to bring equipment to build a new civilization. They brought military equipment. We did not want military confrontation in the new colonies, but we did need the number of women and the colonization equipment to build new civilizations to preserve mankind. We colonized three new planets. The Chinese were sent to Vulcan by themselves. The other planets are populated by the rest of the people in the world who were able to build space ships. The ships from the United States were manned by a hodgepodge of American society which means that there are Chinese Americans, American Indians, Germans, English, Mexicans, French, etc. in roughly the same proportion as the American society. The Europeans brought their own historical nationalities. The Japanese brought Japanese. South Korea brought Koreans. Everyone was required to be fluent in English because English was already spoken around the world and we wanted all the colonists to speak the same language and comingle as they did in America without regard to nationality. All the training, books, movies, references are all in English. There is no point in trying to bring your old nationality with you because none of us will ever return to Earth and our old countries. Note, that all of the countries had some form of democracy versus royalty or dictatorships, except for China. But enough of that. You will learn all of this and more in the learning chair. I'm not sure what SHIP will teach you in electrical engineering, since you are ahead of SHIP in some areas, but once you get your training you will know all of the electronics in our history."

Jane, Hua, had been eating during Craig's statements, "This is excellent. It is as good as anyone's except my mothers. Apparently, SHIP, did not learn her unique methods as they were never written."

SHIP spoke over the sound system, "That is correct. I incorporated every published word, every movie, every television and radio show. I know what all my previous captains have known, but I could not know all the private ideas of mankind."

"SHIP can speak?"

"Yes, I can. I hear everything you say within my structure and I can communicate through communicators when you are outside my structure. I can speak as you would speak to any human and understand much better than humans. I know every language documented on Earth."

Karen asked, "How old are you?"

"In American years I would twenty-nine."

"Do you have any old scars or injuries?"

"Yes, my knee was rebuilt at the end of my gymnastic career."

"Look at your knee now and tell me if it feels any different."

"There is no scar from the surgery and it feels like my knee did back when I was young instead of how it felt yesterday as a well healed artificial knee,"

"That is one of the benefits of teleporting to SHIP. SHIP strips you down to subatomic particles, corrects any injuries and genetic defects and brings you onto SHIP as good as you will be physically at any time in your life. Supposedly we will not age past our prime time in life. That will be a different age for each of us depending upon our genetics. We don't have evidence that this is true yet as the rest of us joined SHIP in our late teens or just over twenty years old and now we are obviously in our twenties. Karen and Craig would be over thirty, I seem to be holding at around twenty-three. We did not realize we had aged at all until a few weeks ago when Craig was reviewing some old SHIP videos of our earliest explorations and realized we looked younger than now."

"You meant my knee is now my old real knee?"

"Yes. Estera wore bottle thick glasses until she transported here and as soon as she got here she had perfect vision without glasses."

"How many girls does Craig have?"

“Seven, counting you. SHIP is equipped for six girls, but four of us are on Pangaea. I’m not sure what happens when we get back to Pangaea. Maybe one of us will have to stay behind if all of us need to travel. We think that maybe Estera will choose to stay on Pangaea so we needed to add another to keep our ratio of six to one. Apparently, there could be many more girls as long as the number on SHIP does not exceed six. SHIP, at first it was only Craig and me, then SHIP changed to fit all six. Can you expand to seven?”

“No. I was designed for the optimum for space travel at six women and one man. It against my programming to allow more than six females and one male.”

Jane asked, “Could you have a different mix, say three women and three men?”

“NO. Throughout history I was designed to take one man as captain. It was only when Craig was determined to escape Earth that I discovered my programming allowed six women to help Craig. That ratio is also demanded by my builders for colonization of space.”

“Why were my people from China allowed to go into space with more men than women?”

Craig answered, “We tried to get them to comply, but the alternative was to deny them survival. We held off on some technology required until the last minute to try to force them into compliance, but we put them on their own separate planet with little likelihood that they would be able to go to or even find the other colonies. You know some of the things we did to guide them once they were on Vulcan.”

Hua (Jane), Dawn, and Karen spent several hours showing Jane around SHIP explaining things as they went and then went to bed. The next morning, as simulated by SHIP, and after breakfast, Hua spent four hours in the learning chair. She came out with a whole new outlook on life from her gained knowledge.

# CHAPTER NINE

# ADJUSTING THE CREW

Twelve weeks later they arrived at Pangaea. SHIP contacted the ones on planet and decided on a meeting place. Since all seven girls could not be on SHIP simultaneously, they had to meet on Pangaea. They were all transported to a well explored and safe mountain ridge near the ocean. Hua Ming was introduced to all of the girls that had not met her and vice versa along with a short explanation of their specialties, how they came to be the crew. Hua knew most of this from her training by SHIP, but actually meeting the other women with an explanation helped understand better.

Craig got down to business, "We explored the destroyed civilizations on the two planets we found. We went to Vulcan and discovered they had a method of cutting and welding Impervium and then found that Jane had built the device that could do it. Vulcan had built small spaceships from the Impervium in the space ships and armor for their vehicles. SHIP did not know that it could be done, but Jane did it. Jane was unhappy on Vulcan and was running out of excuses not be just become one of the women being used by the men. Because she was an Olympic gymnast she knew about the rest of the world and wanted out. She is a good addition."

Estera spoke up, "Good. Craig, girls, I have found someone here that I want to stay with. Craig knows that I was never really happy to be one of his crew. My father traded my survival for his political help in funding the space ships and other technology that had to be built for this colonization. I really like Craig, and am very happy that I was able to come along and do my part up to this point, but I am really just a home body and I have found someone that I can stay with."

Dawn argued, “But, Estera, you will miss out on our future explorations and you will get old.”

“That is my choice. SHIP made me healthy, fixed my near-sightedness, but I want to have children now, not sometime in the far future. All my ancestors died at a normal age. I know that Dawn, Karen, Jill, Sandy actually love Craig in their own ways and are supremely happy being members of SHIP’s crew, but not me, and I am not sure about Shelby. I want a normal life and while Pangaea is not normal I want to stay here and grow with the colony. Yes, I know I will still be sharing my husband with other women, but it will be on a planet, in a house, with sunshine and fresh air. Yes, I know SHIP would keep me healthy and young, but the air on ship is not full of the scents of a planet. SHIP would keep me physically fit, but you don’t have to work at it. I have physical and mental work here with the spreading of earth plant life, and the eventual harvesting of grains and fruits and vegetables. I would miss being around hundreds of people. I have decided to stay here, until I die of old age. Craig, you know that I went with you simply for my survival and scientific interest.”

“Okay Estera. How about you Jill?”

“I will go with you. I stayed here out of duty to the colony, but I have no desire to stay here forever unless you decide to stay here also. I love you more than anyone or anything else so far in my life. The engineering and exploration of space is what I want.”

“Okay, Jill, I appreciate that and I love you and the adventure too. Sandy?”

“You should not have to ask. I met you that one time and then waited for you to come back to me for several years before you did. No way are you getting rid of me.”

“Sandy, you know how I feel about you. I’m glad nothing has changed. Shelby?”

"I won't say I love you, Craig, but I have nothing better going. The earth animals here are doing just fine. I have successfully produced all of the species we brought with us and have done all the research I can do at this time on Pangaea, so I am ready to move on with you and SHIP and all of the women of the crew. I could not be happier to continue the quest wherever it leads."

"Okay. Dawn and Karen have fully committed to me, SHIP, and the quest. So I am left with Hua, or Jane?"

"I am joyous with what I have experienced since coming to the ship. I am not sure what love is, but I find Craig as appealing as anyone I have met. I only hope my nationality does not negatively affect any of our relationships. I have not really experienced what the rest of you have experienced. I learned what SHIP learned on your quest so far and am dumbfounded at what you have accomplished. I already knew most of what SHIP tried to teach me on electronics and electrical engineering. I did learn a lot of technical stuff that I had not been exposed to. Between the capability of SHIP to materialize anything I can imagine or design should keep me providing something worthwhile. I don't know a lot of what we can expect as I only know Vulcan and really have not had personal experience on Pangaea or Pyramid, but that can wait. SHIP taught me what it learned before leaving for those destroyed civilizations on the two planets you found. I am anxious to do my own observations."

"Okay, we have our full crew again. I hope you, Estera, are happy with your decision to stay here."

"I will be happy here. I never felt comfortable as a perpetual adventurer. I don't mind growing old. I will see you on your subsequent visits here. Look at it this way; I think SHIP could rejuvenate me at ten or twenty year periods. If something unforeseen happens to the current crew members, I will come with you again to form a full crew. But if that happens, I hope you will find another replacement like Hua. I'm not counting on rejuvenation by SHIP. As long as you have a full crew, that may not be possible. I don't think SHIP even knows, but regardless, I will be happy growing old if that is what happens."

"Okay, but stay here until we discuss our immediate plans. Everyone, give me your ideas of what we should do next."

Dawn said, "I think we need to take Jane Ming by one of the destroyed civilizations then go on and check on Pyramid. It has been nearly two years since we were there last. We need to have all of us there to check out the pyramid more and see if there is more to learn and to see if the colony there has learned more about anything."

Karen said, "I was the only other one at the destroyed civilizations. I want to know what happened there. We spent a month there, but we really didn't learn much. There has to be more evidence as to what happened. If it was a war between planets, we need to know that. If there is a flaw in our dependence upon Impervium we need to know that too. If it was something more natural or some kind of invasion from outside we need to know. And yes, I agree with Dawn about visiting Pyramid, but I would also like to see Earth again and look for survivors of plant or animal or human life."

Jill said, "I agree with all the above. Count me in."

Shelby said, "I will go along with visiting the destroyed civilizations, but I really want to go to Pyramid."

Sandy said, "I want to see what animal life there is on the destroyed planets and I really want to visit Pyramid to see how their earth livestock program is going. I'm sure they have done okay and they have plenty of indigenous livestock left behind for us by the Martians, but I am the expert on earth livestock. I would also like to see Earth again to see if anything survived. That means we will be gone for over a year. Estera, you will have to take care to stay healthy. Pangaea does have doctors and a hospital set up, but SHIP won't be there to help."

“I know. Aren’t you going to spend a few days showing Jane around Pangaea instead of just what she has learned from SHIP videos?

# CHAPTER TEN

# RETURN TO DESTRY

They spent a week on Pangaea showing Jane around on the surface of the planet using SHIP to transport them from one local to another and then using on planet transporters on the ground. Then they took off on their voyage to the nearest destroyed planet.

Jane was amazed at the differences between Vulcan and Pangaea. Not just the benign nature of the planet itself which seemed like the Eden of the Bible but the makeup of the people. The ratio of women to men was still close to the six to one, but the number of children far surpassed Vulcan. Here on Pangaea the women were all very capable, highly valued members of the larger society with no hint of the near slave character of women on Vulcan. “I want to leave behind my design for the electron beam just in case we don’t get back. The science here may far surpass what I have done.”

“I think that we will be safe on this trip, but we will be gone for many months and we might all be surprised at what the scientists here can do with your invention. Thank you for trusting us.”

“Craig, how could I not trust you? I was somewhat suspicious when I first joined your crew. I was suspicious of the training SHIP gave me. It could have been propaganda, but after seeing the colony working here on Pangaea, I fully understand why you sent the Chinese to Vulcan and wanted to keep them separate from this peaceful planet.”

They arrived at the destroyed planet they called Destry twelve weeks later and spent a week surveying from space. Jane, Hua Chin, had been with the crew for six months, but had still not teleported that many times since she had spent most nights on Pangaea during their visit and the rest of the time stuck on SHIP. This time they concentrated on animal life on the continent with the intact city. Since they were concentrating on warm blooded animals they found more and more of them and had SHIP zoom in on them. They found cattle like animals, deer, dogs or wolf type animals, rodents of various kinds. But the continents with destroyed cities had no warm blooded animals. Apparently the Marians had left and left behind all the animals. They explored the seas between continents and found the number of sea creatures for a healthy sea so what ever happened on land did not affect the oceans. Even the lakes and rivers on the destroyed continents seemed to have water life though in much smaller numbers than the healthy continent.

Sandy provided her theory, “I think that somehow the water life was protected from whatever happened here, but the destroyed continents with little animal life is unhealthy due to lack of carbon dioxide for healthy plant life and consequently there is less food to support water life. The surviving continent has a lot of animal life to keep the plant life comparably healthy which in turn promotes the water life. The ocean life is more constant since it is more dependent upon other ocean life from sea weed to smaller fish, etc.”

Jill chimed in, “I agree with Sandy. Apparently whatever happened on Destry happened in the atmosphere and the water life was protected by the water. We need to find out what happened.”

Dawn said, “I think we should explore the surviving continent more for more archaeological evidence to see if we can find the cause.”

They eventually settled on teleporting to the largest intact city where vehicles were parked near what appeared to be a space port. Sandy stayed on SHIP to study the animal life SHIP had photographed from space. SHIP also followed her directions in directing its sensors where Sandy directed. SHIP, of course, kept in very close contact with the crew on the ground. All of the girls except Hua were teleported to the location with all their equipment. Hua had to be transported without even her clothing with Craig, just as all the women had to be teleported until SHIP learned their exact molecular makeup. Her selected equipment and clothing were waiting when they arrived. Craig said, “Don’t worry, Jane, you have to give up your modesty at first. SHIP has learned the rest of the crew so that they can teleport on their own with all their clothing and equipment. SHIP will learn you after another forty or so teleports so you will be able to teleport without me and we can be clothed during teleportation. Right now, you need to stay close to me in case of trouble so we can teleport back together. SHIP will be in voice contact with all of us and can lock in on all of us except for you if you were alone. I don’t expect any dangers, but we have learned to play it safe. Karen even had to save me a couple of times and vice versa.”

When they rejoined the others already on the ground, Craig stated, “This is the last place where people were before leaving the planet. Because of the concentration of ground vehicles I think we can assume that the people congregated here for an evacuation. I don’t see any other explanation for the concentration of ground vehicles. I think some of them may have been flying vehicles.”

Jane immediately went to one of the vehicles and stated, “This is an electric vehicle, but I find no batteries and see no way that they received their power except what appears to be an antenna on the roof. There is considerable computing power on board so the vehicle should have been mostly self guided, but it does have controls for manual guidance. It probably went to its destination by voice command and then the driver could take over for the final parking or when instructions were unclear. I can see infrared and radar sensors that were probably used for avoiding traffic and people.”

"That is more than what we found." Craig replied.

"Can I just wander around on my own?"

"We don't know what kind of dangers there might be, let Karen follow you around. She is the best warrior among us. If there are any dangerous animals, she will take them out. However, I don't want either of you to get more than one hundred yards of me so I can run to you if you need to be teleported in a hurry."

"What do I do while you are coming to me?"

"Dump your equipment and strip your clothing off. Be quick. I should get to you in ten seconds or so and can teleport to SHIP as soon as you have gotten everything off. I do not know any dangers that Karen cannot take care of, but our experience here is limited."

Dawn, Jill, Craig, and Shelby started looking in vehicles on their way to what they thought of as the space port while Jane led Karen toward what appeared to be an aircraft. "Hey, this vehicle has something left behind. Come see."

The vehicle was probably the fiftieth they had checked and it was sealed up tight with a box like object in the back seat. "Don't touch it until we can get Karen here, she is our official archaeologist now. Don't even touch the doors; we want it to stay sealed until we are ready. Do you have anything in your pack to mark this vehicle until we get back to it with Karen?"

Dawn placed a yellow scarf under one of her tools and left it on the top of an adjacent vehicle. Then they continued their vehicle inspection as they went their separate ways toward the building they assumed was a space port terminal. After another hundred vehicles they came upon some sort of cargo transporter that seemed to be fully loaded with metal boxes. Craig used a communicator to call Karen and Hua to their location. Karen asked, "Is it important? Jane seems to think this a four man aircraft that she is inspecting."

"We have found one passenger vehicle with a box on the back seat and then we found what must be a truck loaded with boxes. You need to see them and help decide what to do."

Sandy was listening in, "Why not have SHIP take a look in the boxes with the sensors?"

"Good, SHIP, do so and give us a report on each box, one by one. Don't look at the next box until I tell you its okay and keep a record of each finding and our comments. Sandy, keep an eye on what you see in the boxes and speak out. And SHIP start with the box furthest away from where we are standing."

Karen and Hua walked up then and Karen said, "Whoa, this may unlock a lot of mystery. They must have had a full space ship and someone decided to bring it all their stuff which they had to leave behind. What's going on now?"

"We're having SHIP use the sensors to look in the furthest top box from us. Would not want to open an unpleasant surprise and Sandy is watching the sensors to see what she sees that SHIP may not report from our viewpoint."

"Good thinking."

"Karen, Sandy thought of it, not me."

"The boxes are made of Impervium, probably for durability and light weight. It appears to contain clothing and foot coverings."

Sandy added, "That would be shoes similar to human. This will tell us a lot about the sizes and shapes of the Martians that lived here."

Apparently the boxes contained personal items from multiple families. Finally, on the bottom level SHIP found something more interesting, a box containing electronic equipment of some kind. Sandy said, "Jack Pot! I think we have a box of electronic equipment. Maybe it has recordings of what happened or at least some pictures of the inhabitants."

“Is there any danger that there may be microbes dangerous to us?”

SHIP replied, “I suggest that I teleport this box here and let me inspect it isolated from the crew.”

“Karen?”

“Go ahead, but I don’t think any of it will be dangerous. Let’s open a box of personal items, right here. It can’t be as dangerous as opening a grave?”

“Okay, Karen, but just how are you going to open a sealed box made of Impervium?”

“How about I use Jane’s electron beam?”

“Jane, can we use your beam to open these boxes?”

“Sure, but SHIP will have to provide a wired power source that produces twenty thousand volts at twenty amps and I will have to operate it.”

“SHIP, can you provide such a power source?”

“Yes, it will be no more difficult than providing a Dodge Viper back when you first became my captain.”

“Beam down what we need. Jane, will you help me set up the wiring and your beam device then I will get out of the way and let you go to work.”

It took less than five minutes to cut one side off the box. “Karen, after I help you unload a box let us back away, just in case. SHIP can decontaminate you during teleportation to the SHIP before we get too close to you.”

“That does not explain why we did not find possessions in any of the buildings we looked in.”

“No, it doesn’t. Maybe this happened centuries ago and their personal items not in this type of box just dried up and blew away.”

"Okay, Karen, go to it, we'll back off for a bit, but keep your com open."

"It appears there clothing is much like ours except there are no zippers on anything, just hooks and buttons. Apparently, zippers and Velcro were an Earth invention in recent years. I would say the average male must have been around six foot four and weighed around two fifty. The female clothing would make them five eight or nine and probably one seventy. There is children's clothing, but all over the board size wise. Men's shoes are all pull on, maybe they never thought of shoe strings and I would guess size fourteen. Women's shoes appear to be around size eleven. Interesting in that none of the shoes have heels. Any ideas? Wait, I have an idea on the shoes. Horses on Earth were much smaller thousands of years ago and these were big people. Buffalo were huge. I suspect that the Martians never learned to ride anything, therefore they never had saddles which meant they had no stirrups, and therefore did not need heels that were largely invented to fit stirrups."

"Sounds about right, Karen. Teleport up to SHIP and verify you were not contaminated in any way from going through that box. Report back if you were clear or warn the rest of us to return to SHIP to remove contamination."

Karen was clear. The rest of them had worked their way through a few thousand vehicles until they were at the spaceport, by the time Karen was back.

Sandy reported in, "All of the boxes on the truck type vehicle were personal clothing except for the box of electronics. SHIP has reported it clear of contamination, but could not identify what kind of electronics it was. It was like nothing like anything SHIP was familiar with and therefore some new kind of electronics developed since SHIP was manufactured."

"We'll look at it when we get back to SHIP."

“Anyone ready to go into the building?”

“Craig, wait. I enter first. I am the warrior and can usually sense danger.”

“Okay, Karen, ladies before gentlemen,”

“Give me at least five minutes before you come in.”

***

They all entered and searched. They found some personal items in the building that apparently remained sealed over the centuries. Shelby, “I found some kind of box with some protuberances”

“Some who?”

“You look. It doesn’t have knobs. None of the things move or else they are frozen, but it looks like it ought to be a radio or camera or something. It is about the size of a cell phone, but there are no controls.”

Jill called out, “I think I found the control room.”

Dawn said, “This looks a lot like the inside of the pyramid.”

Shelby said, “Look, there are seven control stations. Shall we give it a try?”

Craig said, “No electrical power.”

Jane made a suggestion, “You told me during the journey that the vehicles here appeared to operate on an unknown source of electric power. I was inspecting the vehicles outside and the aircraft type vehicle and they seem to have antennas for transmitted power like radio waves. There are thousands of cars out there. Maybe we could roll one separate from the mass and use my beam weapon at lower power settings to try to find the amount of transmitted power they used. If we use too little it does no harm. If we use too much power, we still have thousands of vehicles to test on. If we could get the right power level maybe we could power up this control room and maybe play some recordings."

"What do the rest of you think?"

Jill said, "Great idea. Let's see if we can roll some small vehicles out from the peripheral and you can start out with a low power beam and work your way up to see if you can power a vehicle. Then we could take some smaller electrical devices out and try them at the same power. If that works we could then aim the beam at the antenna on the spaceport to see if we can power up the control room."

Karen said, "Okay, but I suggest that after we have rolled out some vehicles for testing the rest of us could continue our explorations. Craig, you will stay with Jane to transport her back to SHIP in case of trouble and I will stay with the rest of the team."

Shelby said, "I want to do some chemical testing on the waters and soil to see what I can find."

Dawn said, "Karen, I agree. Shelby lets do what Karen said then, we will stay together and watch you do your chemical testing later."

Jane started out transmitting her beam at 10000 volts and the vehicle she aimed at just started smoking. "Craig lets move my beam machine over and try another vehicle at a much lower power setting."

They did. Jane started at 500 volts and increased the power as slowly as she could. The next voltage increment was too much and they smoked another vehicle. "I need to go back to SHIP and get SHIP's help in designing a new control to better increment the power settings."

Craig contacted the rest of the team by communicator while Jane stripped and then the two of them teleported to SHIP, then her equipment was teleported, then the generator that SHIP had teleported down. SHIP and Jane worked on designing and testing for a week. Sandy stayed on SHIP to keep monitoring the sensors and studying animals. Craig and the rest of the team continued exploring. After ten days, Jane proclaimed they were ready for another test of her electron beam. Craig and Jane Ming beamed down followed by her clothing and the equipment. SHIP had built a new beam gun to Jane's designs and created a new generator so that the two pieces of equipment could both be adjusted for infinite power adjustments of transmitted voltage and amperage to try to find the right energy levels for the vehicles and then, hopefully, the other Martian equipment.

They picked another isolated vehicle and started gradually upping the voltage and amperage. They smoked another six vehicles until they accidently found something close to the correct transmitted power and the vehicle powered up. Jane very gradually adjusted the power until she found the point where the vehicle's electronics smoked. "That is the upper limit of voltage and amperage. Anything higher in either range is too much. Both are surprisingly low. I can't believe it would take so little power. Their electronics must be extremely efficient. It is almost like the natural static electricity is providing much of the power and it only takes a little more in the right frequency to initiate the power in the vehicle."

"Why don't I get in the vehicle and see if I can get it to operate?"

"The beam is so low power, it should not hurt you, but let me do some testing before you try it. SHIP, please send me some testing equipment so I can mount it on the vehicle to test electrical potential, voltage, amperage at very low levels and then record those results. I will mount the testers and then turn on the beam to power the vehicle and turn the beam back off and study the testing results."

The testers appeared with sticky glue to hold them in place. Magnets would not work because Impervium is not iron based to be magnetic and Jane did not want anything sticking to the vehicle that could not be easily removed. After attaching the testing equipment to the vehicle she walked back to her electron beam and powered it back up gradually increasing the power to just below where the vehicle electronics would self destruct. Then Craig and Jane returned to SHIP to analyze the results of the testing.

"Well, Craig, I think it is safe to be in the vehicle when I power up the beam. At the highest power setting I used you might feel a very slight shock if you were grounded and touched the outside of the vehicle. At worst, it would be like walking on carpet and touching a ground. A little spark, but only instantaneous and not really painful. I don't plan on using that much power."

Sandy jumped in, "Craig. Forget it. You cannot be replaced on the SHIP. We know from adding Jane when Estera stayed on Pangaea that one of us women could be replaced, but if something happened to you, SHIP would have to search for a new captain and a whole new crew with years of experience lost. I will go first. If there is no harm and I can get anything like operation from the vehicle, you can try it."

"I understand your reasoning, but I would not want to lose you. I connect more with you than anyone else."

"But you have other women, we have no other guy. A new captain would most likely spend years gaining experience then find a new crew. There is not as much choice for SHIP as there was among seven billion people on Earth."

"Okay. The three of us will teleport down and do what you say."

Jane powered up the isolated vehicle and Jane cautiously approached, then touched, and then entered the vehicle. There was no shock. She sat in the seat. The vehicle just sat there.

Jane said through the communicator, "I am going to try some of the controls that appear to be for moving it. I will try to drive it forward about ten feet and … it moved. I did not touch anything."

"It only moved two feet. Maybe it is telepathic? Look at a spot ten feet from where you are and mentally tell the vehicle to go there." The vehicle moved there slowly and stopped.

"I could hardly feel it move, I could see out the windows that it was moving. I'm going to try to go further this time." The vehicle moved more rapidly this time and traveled about fifty yards while Craig exclaimed, "Careful, we don't know how things work yet."

"This is great. I'm going to turn it around and move back to the starting point." Even as she was talking the vehicle did a one eighty and quickly moved back the fifty yards and slowly stopped in nearly its exact starting point. "How did I do?"

"Great, now go back the fifty yards and try to guide it around a fifty yard square to see if you can control its turns." By this time, the whole team was gathered behind the beam watching. The vehicle quickly moved fifty yards straight ahead, turned ninety degrees in a smooth slow turn and started off then stopped.

Sandy said, "What happened?"

Jane replied, "My fault. I did not move the beam to continue providing power."

Jill said, "According to myth, Atlantis transmitted power from their pyramids to cover Earth with static electricity."

"Why don't we all go to the pyramid here? Now that we have a full team of six women and one man, maybe we can gain admission and power it up."

"Great idea, Craig. We'll meet you there." The others were able to teleport there directly. Jane had to strip down and teleport with Craig. Jane's equipment was teleported to SHIP and Jane's clothing was teleported to the pyramid for Jane to get dressed again.

Jane (Hua) said, “Are you sure that stripping in public is necessary to teleport?”

“Yes, Jane, for now. At the rate we are teleporting back and forth, SHIP may be able to do without that in another few weeks.”

“I am not overly self conscious and it is only you and the other five women, but it is still embarrassing and time consuming to have to strip and get dressed every time we teleport. It also ties you to me so you do not have full freedom of movement.”

“We went through this for years. I think SHIP was getting faster by the time I added Sandy to the crew, so maybe with SHIP’s experience it will be even quicker this time.”

Shelby said, “We all went through what you are doing and you have a great body so don’t be embarrassed. Besides, you are so small, maybe SHIP will learn you more quickly than it did the rest of us.”

They walked up to the pyramid in the center of the city and found the large palm sized buttons like on the pyramid on Pyramid. “Okay, Jane, you are new to this. On Pyramid we discovered that the center button only works with a male and the other six buttons only works with women. Everyone pick a button and on three lets all place our hands on them and press gently.”

They stood back as a door opened and the lights came on in the pyramids interior. This pyramid was much larger than the pyramid on the Pyramid planet and had corridors and rooms off the corridors. They worked their way to the center and discovered a large door that required more palm buttons to open. When it opened they exposed huge pipes and mysterious equipment.

Jill postulated, “I think this must harness geothermal energy to operate turbines to generate huge amounts of electricity. When I tap on the pipe they don’t sound hollow, they sound like they are filled. It is probably not water, but probably some unique liquid that changes state with heat and cool. The pipes go up around the inside of the pyramid to cool the liquid then pumped back down to the heat to return as steam to power the generators then around the inside of the pyramid to cool. The square corners of the pyramid slow the gas to help in cooling as it circulates around the pyramid. Now, if we could figure out how to power it up, maybe we could power this whole city, maybe continent. This is the only intact pyramid we have found on Destry.”

“SHIP, can you track us okay in here?”

“Yes. I am locked onto everyone except Hua. You will have to stay with her Craig.”

“Everyone pair up and explored in pairs. Keep talking through your communicators so everyone will know what you are seeing. Karen, why don’t you come with Jane and me? That means there will be three sets of us exploring.”

Eventually Shelby and Sandy reported, “I think we have found the central control room. It is on ground level in the middle of the center of the side furthest from the door. I think if you go to the door and take the second corridor on your right then the first on your left it will lead you to the back then go your left again until you find us.”

The first to arrive did not say much but just looked around at the blank screens and unlabeled controls. One thing evident was the palm sized buttons like on the outer door of the pyramid. Sandy asked, “Well, what do you think?”

Craig answered, “I agree, this must be the main control panel and I see the start buttons.”

“Do you think we will need Jane’s electron beam to power this control room?”

"Maybe not. I can't imagine they would need external power to start up an only power source on the continent. Now, if it uses batteries, maybe they have gone dead in the hundreds or thousands of years since people were here. It could be that we might need the beam to jumpstart the system. We'll see when all seven of us get here. Next question is should we power it up if we could?"

"You mean, how could we turn it off if something bad starts happening?"

"Exactly.

After the whole team arrived Craig continued, "I suspect this is nearly a perpetual motion machine. There is some reason it was shut off before the last person left. Unless it has to do with some limited life of Impervium that we are seeing signs of here on Destry. The problem here is that if we successfully start the system, will it operate properly without blowing up like on the destroyed continents? To start the system all seven us must be here in the inside of the power pyramid furthest from the exit and therefore if it were to blow up it would get us all and we would not be available to help the colonies or continue scientific development. As we discussed earlier, if something happened to me we don't even know what SHIP would do. SHIP was programmed to select new captains in order for humanity to escape Earth, but that is now past tense. Would SHIP just continue to orbit Destry without purpose and direction? Would SHIP go back to Earth, a dead planet, and try to find a captain? Would SHIP wander the universe looking or a captain or would SHIP go to one of our colonies?"

Dawn asked, "SHIP, what would you do if something happened to Craig and he was no longer captain."

SHIP replied, "My programming was to find successive captains to get humanity into space before Earth and to take orders from my captain that succeeds until I am no longer needed."

"Then what?"

"My programming ends there. There is no contingency programmed."

"Could I give you an extension of your programming?"

"As captain that has succeeded in the mission, I would accept whatever you directed."

"Then, I hereby program you to carry on if I were to die by going to Pangaea after I die or otherwise tell you to continue without me and find a new captain to continue the expansion of science and spreading humanity to further planets for the purpose of insuring the survive of humans."

"Accepted, but I can zap you out instantaneously. Unfortunately, I can only teleport one at a time unless you are hugging the other person."

Jane asked, "Can't you just tell SHIP to treat me like the rest of the crew in teleporting me?"

SHIP answered, "No. I must teleport you many times to make sure that I keep your molecular structure separate from other animate and inanimate objects. Watch the movie of the human fly. That was a true story, but took place on Mars before the colony on Earth. The first experiments ended up with clothing imbedded in the human body. The same thing would happen with a moving robot…the parts are moving and it is hard to put things back. If I create something I can teleport it because I know how it was made, how it moves. Humans are breathing, their hearts are beating, and they have small movements. It takes multiple teleportations before I can learn all the individual sub-atomic structure of a human body where I can tell it different than its clothing. I cannot even lock onto you except I can lock onto Craig and teleport whatever he is holding or wearing. There is a small risk in transporting you when Craig is wearing clothing. By the time we get back to Pangaea I should have learned you well enough to teleport alone like I do the rest of the family."

"This is the first time I have heard anyone call us a family."

SHIP replied, "Since the advent of colonizing new planets a colonizing family has been six women and one man. Your family insists on calling the family group a crew, but on Pyramid and Pangaea a family is still six to one."

"You mean Estera quit the family and stayed behind for me just to be one of six women sharing one man?"

"Yes."

"Okay, tell me this, Craig will remain at his prime age for as long as he teleports to and from here so he has a chance of keeping up sexually with the six of us. But in the rest of these large families, the guy will not necessarily keep up with six women."

"It won't take long without birth control for several of his women to get pregnant and then have children. Remember, everyone in the new colonies has a job to do except for health and when the children are very young. It is entirely possible that one six woman family could have twenty pre-school children at home which will require at least two women of each family to stay at home with them. Rather than have one child per year, they will cut back on sex. As they get older they will not want sex as much. The man of the family will be young enough to keep up at first and then won't be called upon as much as he ages. Having so many people in one house with so many women and their children keeps things busy. It's not like the women send all the kids off to school and just sit around all day. Later as the colony grows, the number of men to women will grow to where after a generation, there will be a roughly equal number of young men and women, but there will still be no need for birth control because it takes thousands of years to colonize a planet.

They decided that they did not want to take the risk of starting the power pyramid until they had done some more critical things for the colonies. Craig summed it up, “It’s not like we need the electrical power here. We have no more spaceships for bringing enough people here to start a new colony. We have no more seeds and embryos for starting an earth based echo system. It may be hundreds of years before we had enough excess humanity or anything else for a new planet. SHIP could probably teleport me out of any danger, but can only teleport one at a time and therefore I would lose all of you which would be unacceptable from both a personal standpoint and even a scientific one. We need to learn how to make one of these ourselves, so maybe we should go back to Mars and research their databanks there to try to find the design, but there is no point in powering up the continent.”

They spent a month researching the planet, Destry. They discovered what must have been a crematorium indicating that they practiced cremation. That would explain why they found no burial grounds or human remains. They found tons of possessions left behind, but they were mostly tools. The only clothing items were in the Impervium shipping boxes and sealed vehicles near he spaceport.

Therefore, whatever happened must have been centuries ago for the clothing to rot to dust. The houses and apartments were not sealed. They found rats and insects that could have destroyed any clothing items with microbes taking care of the remainders. They had been hoping for time capsules, but found none, which led them to believe the ancients did not care about preserving history.

They found thousands of small electrical devices which they could power with Jane’s electron beam set to very low power. They had gotten good enough at it they thought they had calculated the output power of the power pyramid. Apparently that had worked by supplementing the natural static electricity to provide power to the civilization. The sheer amount of power must have been astronomical, but the amperage at any one location very low. This would indicate that the ancients had some form of power transistor that could collect static electricity and amplify it for use to operate electronic. The feather weight of Impervium meant that the aircraft itself was inconsequential compared to the load being transported. Possibly they even generated anti-gravity like SHIP. That was very likely considering that is how SHIP worked: repelling gravity one way while attracting in the other direction. It could be that the aircraft worked by making the entire aircraft and its cargo or passengers nearly weightless and then pulled itself toward the destination and repelled against the departure point. In fact, it appeared that the ground vehicles worked in similar fashion using motors only for fine control in parking or very slow motion. They could operate the vehicles there only for very short distance and only when Jane's electron beam was aimed directly at the vehicles. They used a conventional battery powered vehicle supplied by SHIP for short range transport and teleportation for longer distances. The fact that Destry used vehicles at all was probably because of the monumental computing power required to teleport single objects. SHIP could only teleport one person or one item at a time with all of its computing power. Teleporting hundreds of thousands simultaneously would be impossible. SHIP could only teleport line of sight which further limited range for even one at a time. Teleportation of the team meant that SHIP could teleport one of them in the blink of the eye, but seven of them meant seven blinks and each piece of equipment required another blink. Fortunately, equipment could be put in a pack and the entire pack would only take one blink. If each of them had separate equipment set then it took fourteen blinks to move the team. Try it.

"They did find some beam weapons that must have been used for hunting. They were able to try them out by using Jane's beam and found they could be adjusted from stun to vaporize to subatomic particles. If aimed at a cow they could just stun it and carry it away with the cow waking up unhurt a few hours later. They could kill a deer, they only did it once, but the continent seemed to have an excess of deer and probably needed to be hunted. The dogs or wolves or whatever were like neither and dangerous. They fired a weapon at one and it just ceased to exist with hardly a puff of smoke. The weapons could be quite effective if there was planet wide power, but with Jane's little short range beam the weapons were basically worthless. Karen's arrows and guns were much more effective for the family. They did not find any large weapons or caches of weapons so there did not appear to be any need for a military. That implied that the inhabitants had no fear of war, but believed in owning their own weapons. Whether they feared their neighbors or just used the weapons for hunting the family could not know. They did not find any evidence of a military or even a police force.

Karen said, "The American Indian nations used to be that way. They had very little crime because everyone was armed and their tribe distributed punishment in accordance with the crime. If a life was threatened, self defense was accepted. If there was any doubt as to whether it was self defense the tribal council worked much like your judge and jury and determined whether someone was guilty. This happened within hours of the crime rather than dragging out for years like in the modern court systems. We have not considered much about the legal system in the colonies. Vulcan is a dictatorship, but the other two colonies are pretty much like the Wild West except that everyone is highly educated and just trying to make their colony survive. However, we are bound to have crimes being committed. How will they be handled?"

Craig replied, "Good question. We are more like the American Indians. We have no jails or prison system. I think when we get to Pyramid we need to address this."

Dawn said, "If we go to Pyramid and spend considerable time there, then go to Earth and spend a few weeks observing there, then Mars and then return to Pangaea, we will probably be gone nearly a year. It may have taken care of itself by then."

Sandy said, "Shelby, Estera, Jill, and I spent months on Pangaea while you were here and going to Vulcan. The only things I know about were some women moving to different family units because they wanted to and some jealousy. There were men who found themselves losing their family and other men with overly large families. The people got together to ameliorate that. The consensus created a sort of law that said a woman could not change families unless she could find a woman to move to her husband to keep the family units leveled out as close to the standard six to one as possible. There were a few men and women that died due to accidents or infections so not all families maintain the six to one ratio, but that is the community making laws as required. There were other incidents where someone "borrowed" someone else's tools and did not return them. The colony has a very good inventory of what each person was issued, but not everything had a serial number. A council of neighbors just ordered the tool returned to the person that the council thought really owned it. In some cases people traded tools or food for other things and then were unhappy over the trade and the small neighborhood councils have listened to the issue and made a decision based upon whether the trade was honest. So it appears Pangaea is following the American Indian tradition. This works well when the population is low, but what about when there are millions of people and crime occurs between strangers?"

Dawn said, "I suspect we will find the same thing on Pyramid, but we can discuss this further when we arrive."

Craig announced, "I think it is time to leave for Pyramid. We have searched Destry for a month now. It will take a couple of months to get to Pyramid and we can further analyze what we have found and what SHIP has recorded through its sensors while we are en route. Does anyone have any other suggestions?"

Dawn said, "I suggest that we take the direct route there, but rather than go at max speed lets record more of space on the way by going slightly slower to allow SHIP to look for other potential civilization or livable planets we may have missed. We will be on an entirely new course not traveled."

"Sounds good, but I suggest we still go fairly fast. Any one object to taking three months instead of two?"

"Okay, SHIP, all ahead two thirds for Pyramid on a direct route."

Do you mean two thirds of maximum speed or two thirds of the speed I used when leaving Pyramid?

"The latter, but what is your maximum speed?"

"I do not know, but at maximum speed I cannot sense even solar systems, let alone planets and cannot fly to a point. In theory I could cross the galaxy in a few months, but the cruise to a known planet might take months from where I slowed down. Intergalactic ships come with a map that allows better navigation from point to point. I do not know, I only have references to a galactic map, not the map. If I had such a map I could create that three dimensional map like Dawn had me make in the living area. That galactic map might show the whole Milky Way galaxy or might show the Milky Way as one point on a larger map, or zoom in on a portion of the Milky Way. A human navigator could point out the current location on that map and the desired location and I could go maximum speed."

"Okay, SHIP, why couldn't you display that map you showed for Dawn and have her point out our location here at Destry and then Pyramid and have you zoom there in days instead of taking months?"

“Because the map only shows approximate locations with a lot of unexplored space in between. With an accurate map I could guide around solar systems to prevent impact with a massive sun. As we travel more in this area of the galaxy I can refine my own map, but we did not travel a straight path from here to Pyramid before so there are likely large objects to guide around.”

“Okay, make your map to Pyramid and take us there as fast as prudently possible while looking for other habitable planets.”

“It will take approximately eleven weeks.”

# CHAPTER ELEVEN

# RETURN TO PYRAMID

During the eleven week voyage to Pyramid, the family had considerable time to work on their research. They also had time for extraneous ventures on the ship to relieve the boredom. Except for Hua Ming, Jane, space travel was a common place event. The image of stars and solar systems streaming past the windows was, “old hat”.

Jane worked on her electron beam to build a smaller device to power a single Destry vehicle or hand held device for their research. The reduced power and size created while on Destry was used by several of the crew. Jane and SHIP made it smaller and more self contained so that the power supply and beam were smaller to be able to be used by a single person. Rather than a large potentially powerful power supply, it was downsized to the size of an Earth lantern battery that operated as a fuel cell to generate small amounts of power. The beam itself was downsized from a tripod mounted heavy cannon looking beam to where the combination was more like a half size leaf blower.

Jane worked with SHIP to design a much more powerful version that could cut Impervium from longer distance or warp large panels to make Impervium malleable. The larger version had to be strictly design only and theory because it could not be used on the SHIP. Jane’s phenomenal understanding of electrical engineering and SHIP’s knowledge of all the electrical engineering in history made this very possible. What they had learned by building the smaller power one was just reversed to theoretically create one far more powerful. “Craig, I am concerned that I am creating a super weapon. If I were to keep going with this increase in power I could end up designing something that could vaporize a spaceship or a building.”

"That is not what I asked for, but I think you should look at two or more levels of power increase: what you started on, a method of forming Impervium and two, a weapon. Vulcan was attempting to develop weapon zed space ships to attack the colonies. Maybe we need a weapon. Having a weapon and using it are two different things."

"It's not that I don't trust you, but it makes me feel like Einstein or Oppenheimer developing the nuclear bomb. I see your point. That is what Vulcan wanted from me."

"How sure are you that Vulcan cannot use your work there to develop one of their own."

"I held the design very close and you had SHIP take the beam weapon when we left Vulcan. However, there is always the possibility that I was spied on. I tried to prevent any spying, but I suppose it could have happened. Okay, I will work on both up-powered beams."

Sandy spent much of her time looking at the life found on Destry. "I still don't understand what happened to the animal life on the dead continents on Destry. On the live continent, there were cattle, not like on Earth, but cattle, deer, wolf/dog predators, mice or rats, birds, and other animals, plant life, insects. The carbon dioxide content of the air was very low everywhere. The dead continents were especially low on carbon dioxide which would explain the anemic plant life there. The anemic plant life would not support heavy insect life which would reduce pollination of plants which is a cycle of less and less plants. The lack of insects and healthy plants would explain a dramatically reduced bird life. It's like something killed off all the animals on the dead continents and that led to plant life barely surviving. The one live continent was providing some world wide carbon dioxide which is probably the only thing keeping the anemic plant life alive on the dead continents."

"Maybe, humans could start some plant fires to increase the carbon dioxide and import animals from the live continent and get the cycle going the right direction."

"I think that would help, but it would take more than just us. If we were to colonize Destry at some point, we would likely be using fossil fuel which would help the carbon dioxide of the whole planet. The colonist themselves and the animals they bring with them would help. More carbon dioxide would make all the plants healthier that depend on carbon dioxide. Plants need it just like animals need oxygen."

"Okay, Sandy, keep working on it. See if you and SHIP can do two things. One is to figure out exactly what colonists would need to do to make and keep the planet healthy. Second, see if you can work backward to figure out when the disaster happened. Maybe you can confer with Shelby on the chemistry and Jill on the petrochemical side."

"I did not have time to look for searching for oil on Destry, but any old planet should have animal and plant life that user millennia should have turned to oil. I did not see any evidence that the inhabitants of Destry used petrochemicals or burned anything, period." Jill went on to add, "However, during the voyage to Pyramid, I have been looking at SHIP's sensor data and found areas on all the continents that are so high in petrochemicals that if we were searching for them, we should find exposed coal deposits and even pools of crude oil on the surface. Without drilling test wells there is no way to determine reserves, but there seems to be plenty for an easy start."

"All three continents seem to be equally endowed with petrochemicals," SHIP intoned.

"I don't know what made you think of Chemistry," Shelby said, "but I took soil and water samples everywhere we went. The destroyed continents had very little nutrition in the soil, I did find a few soil dwelling animals like roly polys or pill bugs, and traces indicating earth worms, but other than that, there was no trace of animal life remains. The soil and water has almost zero acid. Destry could use a little acid rain. There were traces of many minerals in the water that just made the continents leaning more to lye instead of acid. Plants cannot live in an acid free soil. The water is grossly out of balance for fish. I agree that burning some fossil fuels might do the planet worlds of good."

"Well, keep at it," said Craig as he walked off toward where Karen was staring at some artifacts they had teleported up from Destry. "Karen, what are you staring at so intently?"

"Oh. Hi, Craig. Dawn and I were studying some of the things we brought from Destry. The clothing we found looks to be very human like. It has a lot of cotton, wood fibers, and animal skins. There are no artificial materials at all. That is probably because they never used petroleum for anything other than crude lubrication that they got from surface deposits of crude oil. Shelby also identified fish oil for lubrication. Therefore, they never refined oil as we did on Earth."

"The most exciting thing we have found is that we could use Jane's low power beam to power what turned out to be video communication equipment. They are very much like our old cell phones on Earth except they had no batteries. They were apparently powered by static electricity in the air supplemented by their power pyramids. Consequently we have been able to capture moving images with sound from some of the devices so we can see what the people looked like and sounded."

"Stop right there," Craig said. "Girls, lets gather in the living area. SHIP, put some of these together in a logical montage of video and sound so we can all watch together."

SHIP compiled only ten minutes, but it was enough to show the people and SHIP could roughly translate the sound to English so they could all understand the language. "I apologize for the translation. The language is basically Atlantean from Earth, but there were terms and words I could not understand. Apparently the language has changed since I was manufactured."

"That's understandable, all languages change with time and temporary fads. New inventions require new words. Unfortunately, these were all family pictures. We get no science from them," Dawn said.

"Not so, Dawn," Karen said. "We see what the people look like. They look like Cro-Magnons from Earth, but taller. Cro-Magnons had a larger brain than humans, but the men were only five six and the women five foot to five one. These Cro-Magnons, if that's what they were have less hair and are much taller than the average human."

"I would like to propose a theory," said Dawn. "I think that maybe the Cro-Magnons we found on Earth were refugees from Atlantis that got left behind. Without their power pyramids they lost their transportation, tools, and weapons. They could no longer farm and had to resort to primitive hunting with stone spear tips and berry gathering. The malnutrition caused their growth to stunt. Their superior strength was due to surviving for a few thousand years without technology living as cavemen. The Atlanteans that went to the stars were originally Martians that had millennia living with less gravity on Mars before moving to Earth. Therefore, the Cro-Magnons we find on Destry were never faced thousands of years of malnutrition and were not stunted."

"Sounds logical to me," said Craig. "I would add that the Neanderthals were displaced by Cro-Magnon and eventually the Neanderthals died out because they could not compete."

"I think you have done my job for me," remarked Sandy. "As a zoologist, I would have to agree with Dawn. Unless we find out differently, let's assume this is absolutely correct."

Jill changed the subject, "Don't say we didn't learn about their science. We saw vehicles moving on roadways in the background of the videos. I even saw some airships flying and a couple that appeared to just float in place like helicopters but without helicopter blades. That would indicate antigravity like SHIP uses. We have seen that their communicators worked like our old cell phones did on Earth but with transmitted energy. We saw no antennas on Destry so I would think that they may have used disturbances in the transmitted static electricity to send complex video and voice signals, which would indicate that either these little devices had mammoth computing power or that some central computer took the disturbances in the energy from the pyramids and retransmitted the appropriate signal to another communicator. Whooo. That is stretching things a little, isn't it?"

"Jill, I think you may be right, but we may never know. What do you think Craig?"

"Dawn, I would agree with you. Jill may very well be right, but we are a long way from proving it. Maybe in a few years we will know. We should always be trying to expand our knowledge and experience."

***

It took three months to travel to Pyramid from Destry. Craig was thinking that it was no wonder some of the early captains of SHIPS went nuts having the knowledge of SHIP and no one to tell about it. The first time a captain told someone that was not a crew member; he would no longer be captain and just look like a nut case. Probably some were considered nuts and were just trying to tell the facts. There is no way that I could have spent these months in space with just SHIP to talk with By having a crew of six women things could not be better. I have six different views of everything with their six different backgrounds and education. With just one man and woman there would be the danger of getting into an argument and having nowhere to get away. With one man and two women, there would likely be serious jealousy. With three women one could end up feeling left out. After years with a seven person crew or family, SHIP had been able to literally grow larger and had created separate bedrooms for each of them, each with its own closest and other storage and a bathroom for each bedroom. Each bedroom also had its own small work area with desk and computer monitor so each of them could watch recorded Earth shows or do their own research on SHIP's data banks.

Now they were approaching Pyramid. Shelby had been distant during this trip. I feel differently about each of the six women in my little family. What is love?

I am definitely in lust with Sandy from the first time there on the beach.

I am comfortable with Dawn and we were together years longer than any of the others, but I think she would agree, crew members with benefits. I think I need Dawn's intellect most and I think Dawn puts science above love.

Dawn's old roommate, Jill, lusts after me more than I her. She is like Dawn in putting science above love, but she is far more physical. Jill was a tease before SHIP and now she tears my clothes off and still manages to hold us both off for over an hour of lovemaking.

Hua Ming, or Jane, she uses me to get off, and has no real concern for me other than gratitude for her escape from Vulcan.

Estero, that Jane replaced, was always a political convenience for me and survival for her. Her botany background was essential in getting the colonies established, but less important now. I am glad she found a man she really wants to be with for more than survival. Normal marriage is not something for this generation of six women to each man, but she did found someone to share with on a planet. She has all genetic defects that were fixed by SHIP so no reason she can't have the children she wants. Good luck Estero.

Karen doesn't fit any normal mold. She is like the fictional Amazon woman. She is tall, strong, and more aloof than any of the rest. She is protective of all of them, but particularly protective of me. Of course no one knows what would happen if I were to die and left SHIP with six women and no captain. Karen and I don't get much private time together, but powerful. I feel like a kept man with her. I could fall in love with her if it were just the two of us and if I had been limited to just one wife; Karen might be the best one to get stuck for years with alone on a spaceship. She is intuitive to the point of being psychic. She pretends to be above it all, but really feeling for everyone else; a problem with being psychic.

I wonder how the women feel about all this. Generally, women want one man. Is that true? Are women more jealous than men? By being permanently in my mid twenties and supremely healthy, are I keeping all of them physically satisfied? All I can do is try and looking out for their concerns. Shelby is the one that seems listless now. After years of flitting around space and after her time grounded on Pangaea I think she is getting the desire for a more grounded existence. "Shelby, let's go in the dining room, I think there is something we need to discuss." Shelby followed him in and took a seat. There were always people in the control and living rooms. But bedrooms were too intimate for serious discussion considering the physical relationships between Craig and each woman. That left only the dining room for private non-sexual discussions.

"Shelby, I sense that you are no longer happy with our relationship."

"That's not true. Well, that's not true either. Now that we have established the colonies and have identified all the things necessary for success, my chemistry knowledge doesn't seem that important anymore. I found myself envying the women with more permanent relationships on planet. No, I know that I can live forever on SHIP, but I like the freedom to just walk around outdoors on my own. I like the hundreds of opportunities for interpersonal relationships on planet. Here, we are all cooped up together. Dawn and you have a special relationship of having started all this and the continuing science advances and exploration. Karen is attached to you like a bodyguard, where you go, she goes first to make sure you are safe. She explored on her own before joining the family so now she has a whole universe to explore and since she never had more interpersonal relationships, she doesn't mind the limitations of SHIP. The only thing she misses is being outdoors exploring new planets, but you have been doing that. Jill is thoroughly immersed in science and where else but SHIP could she do what she does. Besides, she may not love you, but she is attached to you and Dawn was her friend before SHIP. Jane is new to you, to sex, to SHIP and never had interpersonal relationships any closer than now. Sandy, need I say more. She waited years for you to come back and add her to your family."

"So, you are a little like Estera, you are looking forward to settling down somewhere?"

"Yes, but I am not going to desert you. You need the magnificent seven for the pyramids and other secrets you might find in the future. When we get to Pyramid, could I look for a suitable replacement for myself?"

"Would you want to stay on Pyramid versus Pangaea?"

"Yes, I think so. I passed on my chemical engineering learning from SHIP to the other chemists on Pangaea. I think that Pyramid could use my knowledge. I think I will like the more permanent structures and benign wildlife on Pyramid. I think I could find the interpersonal relations I miss, and maybe find a husband to share. I am not unhappy with you Craig, and I love the other girls on SHIP, but I want more relationships with a wider assortment of people and a more civilized colony that I think we will find on Pyramid."

"You have to admit that you may like me a lot, but I am not one of your favorites. The main use you have for me from this point on is to maintain the six to one formula for unlocking Martian systems you may come across. I don't believe that I am, by any means, irreplaceable. I think if you try you can find a suitable replacement for me on Pyramid."

"I understand, Shelby. I do not want to interfere with what you want. Just be warned, like Estera, you will no longer be ageless. If you get sick or hurt, SHIP will not be there to renew you through teleportation back to your genetic makeup. You can help me find a replacement."

"Do you mind if I put up the idea of replacing you to the rest of the family and get ideas for a new skill to add?"

"That would be a great idea, Craig, but I wanted to approach you first. Thank you for recognizing my concern and approaching me before I had to come to you."

They decided that they would all keep their eye out for a computer scientist that would maybe fit in their family. There were a number of obstacles to finding someone. All of the colonists were carefully selected to be in the top of their fields, but the fields were everything from mechanics to theoretical scientists in numerous fields. It had to be a female computer scientist capable of thinking beyond current theory. She had to be willing to forever leave her current family situation. The odds of finding such a female without children to leave behind were slim since one objective of the colonies was to reproduce quickly. Then that female had to be willing to commit to a potentially forever relationship with the SHIP family and had to do it with Craig to get on the SHIP and then be willing to be teleported numerous times without any clothing before SHIP could teleport her with her clothing and equipment. To make it more complicated, this woman had to be compatible with the other five women.

Shelby lamented, “If we are not successful, I will stay with Craig and the family. I know you need the seven and I have to put mankind ahead of my personal wants. I don’t really need to be grounded on a planet. I have already spent years on SHIP, but I have just not been as adjusted as the rest of you. Sorry, Jane, I know you are still very new to all this and cannot understand.”

“No, I do not understand. I was grounded on Vulcan since the start of the colony and trapped on the spaceship for many months getting there. None of the rest of you know what it would be like to be traded off to multiple men on a daily basis. I escaped most of that, but most of the women on Vulcan are little more than harlots. I was only able to maintain my status through my intellect and that was running short. Another few weeks and I would have begun to be traded to multiple men like the other women on Vulcan and expected to have many children without knowing who the father was. Can you imagine? Even a harlot can practice some form of birth control, but not on Vulcan. I am so happy to be here on SHIP forever I can hardly suppress my joy. Not only that but I have such unlimited resources to develop new devices in my field. I hope I never run out of new ideas and opportunities to develop new things. No, Shelby, I cannot understand why you would give up Craig, and SHIP, and essentially eternal life and health for a short lived ground person.”

# CHAPTER TWELVE

# PYRAMID

As usual, before landing on a planet, they orbited for a couple of days to survey the general conditions or in this case changes since the last visit. Each of the family or crew observed SHIP's sensors for the information relating to their specialties. It had been four years since their last visit.

Sandy was pleasantly surprised to see that the Earth animals they had imported were thriving on Pyramid while the indigenous animals left behind by the Martians for the feeding of the people of Earth were all doing just fine also.

Jill could see that the coal mines were in full swing, but the petroleum fields were far behind the fields on Pangaea. "The problem is probably the lack of pipelines to distribute the petroleum reserves. We will have to stay here to help get their plastics production going better and get them started on the pipelines to distribute the petroleum."

Shelby was interested in the air quality and found nothing to worry about. "There aren't enough people and industry to have any impact at all. The air and water has not changed since we found Pyramid. Karen, would you and Dawn join me in the dining room for a bit. There is nothing much for you to do on Pyramid, but I could use some help."

When Karen and Dawn had followed Shelby into the dining room she stated her case, "If I am to stay, I need help finding another crew member. I am going to be busy helping Jill in the development of plastics from the petroleum, water treatment plants, and I won't have time to look for my replacement. Could you use your time trying to find a computer expert that might fill the needs we described?"

"I don't see any point in Archeology here; the Martians left this planet prepared for our use whenever we managed to escape Earth, if we did. I don't see any issues with astronomy or astral navigation while here on planet. Dawn and I should be able to concentrate on that."

"Of course, Karen. I can use SHIP to inspect the population to look for potentials and you can find them and visit with them on the surface of Pyramid. When we both agree from a theoretical point of view, then I will join you and we can interview the candidates."

"Good point, Dawn. I am anxious to get on the planet surface and some freedom from these walls and you are much better with SHIP data. How long are we going to allow the spaceships' computers to keep track of the details on everyone on planet? That is a terrible invasion of privacy."

"Don't think there wasn't plenty of that on Earth. A plain Jane secret level security clearance was nothing more than a data search. There is no reason not to let Jane come along with you. You can be intimidating alone. Taking little Jane with you gives you a Mutt and Jeff appearance that should remove any threat of your physique. Face it; you are like the warrior queen while Jane is more like a little girl in appearance."

"Okay, give me some leads and Jane and I will go lead on the ground."

Dawn spent her time in the databases of the people of Pyramid looking for a candidate to join the SHIP. Jane spent her time just observing out the windows. Karen spent the time with Jane to explain what they had found when the first came to Pyramid before colonization.

Craig had spent his time observing the other continents of Pyramid and inspecting the improvements made to the stone buildings of the city, surrounding fields and orchards, and the clearing of the roads prepared by the Martians. After a week of observation, "I think it is time to go down to the city and see if anyone remembers us. Then tomorrow, I want all of us to visit the Pyramid again and see if there is anything we missed before that we can learn this time around."

They appeared near the Pyramid and the planet government building. They were immediately met by the planet leaders and welcomed. The colonists were doing very well. They had restored the orchards left for them, cleared fields and planted Earth seeds and plantings brought from Earth. Their animal husbandry program for the embryonic animals from Earth had gone well. They lost fifty percent of the potential numbers of animals in test tubes or shortly after, but the surviving animals were doing well. They had been eating the archaic cattle brought there by the Martians. They had tripled the number of people on the planet through childbirth. The number of male children exceeded the number of female children so the next generation would be close to the fifty fifty they had left on Earth. The children were highly intelligent because their parents had been carefully selected for the colony and were being introduced to books and educational videos at a very early age. The oldest children were only four. They had lost a few adults due to unwise explorations where they had fallen off cliffs, drowned in rivers and the oceans. There had been a few accidents in the mines and building.

They had enclosed the stone buildings with wooden doors and window shutters. They had managed to make some glass using the educational materials in the spaceships computers. The spaceships were being used as daycare centers, hospitals, and some meetings. They were using parts of the spaceship for storage. Due to the unlimited nuclear power, their spare meat was frozen in some rooms; other rooms were used to store grain, fruits and vegetables. In addition they had started electrification of the city. The outlying spaceships were used largely as living quarters.

Dawn had found a computer expert that had had a miscarriage which resulted in her being unable to have children. There were thirty-six women that more or less lived together rather than living in the family units with six women to one man. Most of the women were not in mourning over child loss, but just felt that they did not fit in family units where five women were having children and one woman not having children. There were another hundred or so women that stayed in their family unit taking care of the children and houses while the rest of the family worked in the factories or fields. The computer expert was a brunette with long hair, a slender face and quite attractive. Physically she was very fit if a little on the slim side being five foot five but weighing only one hundred fifteen pounds. She had a nice figure, but just didn't fit in with the rest of the family she had been assigned to and wanted to spend her time on computer programming. Dawn sent Karen and Jane to meet her.

Karen knocked on the door where Adriana DeCamp lived with five other women without children. Karen had seen her photos and when it was another woman inquired, "Is Adriana home?"

"Yes, who should I say is calling?"

"Just tell her we are from the ship that is visiting and we would like to speak with her outdoors."

Adriana came to the door and introduced herself, "Hello, I'm Adriana and I was told you were from the visiting spaceship."

"Yes, we are. I am Karen Martin and this is Hua Ming, but we call her Jane because it is easier. Is there somewhere nearby where we could talk with you alone. I think you may be the one that may be able to help us."

"How could I help? I met you in one of the classes on Earth before the escape. You and the other leaders seem to be fully in charge and knowledgeable about everything."

"It is a long story. We really need to go somewhere we can have a long drawn out discussion to fully describe the whole situation and not talk in front of your housemates."

"Okay, there is a community room on the next block that we use for group dinners, dances, and church services. Follow me." When they got there and closed the door, "Okay, we have no locks on any doors here on Pyramid, but this is as private as we get and this room has electric lighting since it is getting dark. What is this big secret?"

"Jane here is a new crew member from a planet we colonized but you never heard of, Vulcan. The people from China were all sent to that planet to keep them isolated from the other colonies because the Chinese loaded their ships with military equipment expecting to rule the other colonists. Jane here is one of them, but like most of the Chinese, innocent. She can go into more detail about Vulcan at a later time, but she had made some major breakthrough that we needed and she was an electrical engineer that we did not have in our original crew. Craig, our captain chose to bring her to our ship. When we got to another planet you have not heard of to pick up some crew members we had left with the colonists there, one of our original crew, the botanist, decided she wanted to just stay there on the planet rather than continue exploring space. Jane became her permanent replacement."

"Now, our chemist, one of our original crew that rejoined the crew on that third planet has decided she wants to stay here on Pyramid. Now that we have found three planets suitable for colonization and established colonies on them, we do not need a chemist as bad as we need a computer expert. Our computer on the ship is amazing, but it cannot think for itself, every statement is based upon its historical data. We need someone that can think of something new. Our research on you shows you might be the one."

Jane interrupted, “We know you are not a member of a family and think you cannot have children so have isolated yourself to do your research.”

“That’s private. How do you know this?”

“The spaceships databases keep track of everything on all the colonists on the planet and our ship can pull up that data for our study. Your name popped out as a possible replacement.”

“You have mentioned a total of three planets. Is that it?”

“We have colonized a total of three planets. We have found two other planets that are now deserted but were once populated by a race of humanoids more advanced than we are that traveled there eons ago. We don’t know where they disappeared to. One planet has three continents. Two have had all their buildings and vehicles destroyed, the third continent was deserted intact. They were apparently evacuated or decided to go elsewhere. We don’t know why they left, how they left, or where they went. Jane has not seen those two planets, but now after arriving on Pyramid, she has visited all of the Earth colonies.”

“That is amazing that there are three earthlike planets that we have colonized. We were wondering why there were so few American’s here since America was the driving force. What happened to Earth anyway?”

Karen answered, “It was utterly destroyed for life as far as we know. We are going there next, then to Mars.”

“And you are asking me to go too? Why Mars?”

“Yes, we are asking you to go, but there are numerous issues we have to address. We are going to Mars because it was Martians that created the things on Pyramid for a colony from Earth. It was thousands and thousands of years ago, but Martians came to Earth before Mars lost its atmosphere and created the mythical Atlantis on Earth. Then thousands of years ago, many of the Martians from Atlantis left Earth for the stars because their city state was going to be destroyed. They knew about the interstellar body coming to destroy Earth and created Pyramid as a destination for Earthmen to follow them into the stars. We are going to be studying the pyramid here more, helping this planet with some technology that we have perfected on that third planet we call Pangaea and then eventually going to Mars to try to find out more about them."

"Count me in."

"The current crew has been together for years. Craig and Dawn have been together for ten years on SHIP. We would be asking you to leave Pyramid for many years, potentially never living here again."

"That would not bother me. I came here willingly and expected to have two or more children by now, but you must know what happened to me from the database and that I can no longer have children. As a result, I have removed myself from the families. I still have friends and am friends with my original, but he now has six healthy wives."

"Yes, we know what happened, but you are still healthy. That could be a benefit of joining the crew of SHIP."

"Why do keep saying SHIP instead of just the ship?"

"SHIP is an amazing almost living being. It is unlike your spaceship. It is incredibly faster and as intelligent as any computer could ever be. You will find out about SHIP after you join the crew."

"Where do I sign up?"

"You have to be interviewed by the rest of our family and Craig has to accept you as Shelby's replacement."

"How many people are on this ship of yours?"

"Just seven. Craig, the male captain and six women. There is more than just child bearing for that number. You know that your pyramid requires a full family unit to open up. The ratio was set up by the Martians. We don't know much about the Martians. Maybe they had a different birthrate than humans. Where Earthmen tend to have slightly more boys than girls, maybe the Martians average six girls for each male child? Maybe Earthmen have more boys because for thousands of years we have been warlike and there was a genetic need for more men to fight those wars. I don't know?"

"So you are saying that I would have to become a member of your family with Craig as a shared husband?"

"Yes. Are you willing?"

"Yes, maybe. I have seen pictures of Craig years ago when he was maybe twenty-two and while not a hunk, he was acceptable. If he is the brains behind the colonization, yes, I think so. I have not been with a man in over three years since my miscarriage. If I just had listened to the doctor and not tried to climb that mountain, I would not have fallen, and would not have lost my baby and my ability to bear children."

"When can I meet Craig and the rest of the family?"

"Here, take this communicator and we will contact you and give you a meeting place. Maybe right here sometime tomorrow. How do you feel about not having clothing?"

"Are you telling me your ship is nudist?"

"No, not at all. It is a matter of transportation to and from SHIP. We don't have a shuttle, we teleport to and from SHIP. For the first thirty or forty times you teleport you have to teleport with Craig without clothing so that SHIP can isolate your sub molecular makeup. Later, SHIP can figure out which is you and which is something else and you will be able to teleport fully clothed with equipment. Jane here is still too new for that, but the rest of our family can teleport with everything we can wear or carry."

The family had been given a guest house to stay in rather than return to SHIP. They had been loaned three battery powered Jeep type vehicles. For distance travel they would have to switch vehicles every one hundred fifty miles. Dawn and Jill decided to stay out of it completely and leave the addition of a new family member to Karen and Craig. Jane stayed at the house too, but Shelby wanted to meet her potential replacement and Karen came along to guide Craig to the home of Adriana. They would drive back to their temporary house on Pyramid if Craig and Adriana hit it off because Craig would take Adriana to SHIP and eventually back to their temporary house.

It was dark when they arrived at the community building for Adriana's neighborhood. Adriana was there with the lights on. Many of the houses were still waiting for electrification because their copper wire factory was still only four years old and when you have a whole continent to wire for electricity it would take years yet. Consequently, neighborhoods had initially just had some central meeting rooms wired for electricity. Their only heat in their homes was using coal mined away from the city and trucked in and distributed. The electricity for the planet was supplied from their nuclear powered spaceships, so wiring for electricity was complicated. The charging stations for their vehicles had been wired first so the spaceships were wired together in a network. Factories, the mines, the petroleum fields, and the common meeting rooms like the one where Adriana was were wired first. Eventually, everything would be wired into the network.

Karen went first since she had already met Adriana. "Good evening Adriana. This is Shelby who you may be replacing and this is Craig, SHIP's captain and our male representative to our little family on SHIP."

"Glad to meet both of you. You and Karen here and the rest of your crew were responsible for the colonization of space, which makes you next to gods to most of us."

"We are far from gods. You will understand everything in a few days with us. It was not a matter of any godlike behavior. It was luck, being in the right place at the right time, and ancient ancestors that created SHIP", Craig replied.

Shelby added, "You will be amazed at SHIP, but SHIP has limitations. Craig is too modest, the technology may have come mainly from SHIP, but Craig is the one that had the ideas outside SHIP's capability and brought the crew together. It is rather amazing how he seemed to select all the right people to supply the expertise needed and that we would all be compatible. You are the first not personally selected by Craig. He has final approval, but Dawn, his first crew member found you by searching the data that SHIP gleaned from your spaceship computers here on Pyramid. Then Karen and Jane interviewed you first. Before this Craig had to do all the searching out of his crew and then convincing us to join him in this quest for mankind."

"I am ready to join the crew, but I have several things I have to tell you and some questions."

"Okay, you tell or ask us first."

"I must tell you to start that I can no longer have children. I was exploring when I was pregnant here on Pyramid and fell down the side of a mountain losing my child and my capability to have children. I have a number of scars from broken bones where you can't see them and will probably have the resultant arthritis when I am older. I am now thirty-one years old so in another ten years I may start having physical limitations."

Craig laughed, "That will not be a problem if you join the SHIP's crew."

Adriana did not understand that teleporting to SHIP would eliminate any physical problems and even correct any genetic problems she might have. She continued on with a question on her face, "Okay….. If you say so. Shelby, why do you want to leave the ship after nearly thirteen years? I must say you don't look like you are over thirty. I would place your age at early twenties."

"When we were exploring the near galaxy for new planets and throughout the effort of establishing the colonies all of us were immersed in the mission. Now that we have established successful colonies on three separate planets, mankind is safe from extinction. The colonies are far enough apart that a supernova of one of the stars could not take out more than one colony. That part of the mission is over. Mostly, I have been very happy. I am still happy. However, last year I spent nearly a year on Pangaea doing routine things and being with hundreds of people instead of isolated on SHIP. I mean our little family is very happy together, but now that the pressure is off, I would like to settle down on a planet. I left Pangaea to come along because having a six woman one man crew on SHIP is far more important than any other planet unit on any colony so I simply could not leave them short a person."

"Jane replacing Estera and her staying behind there made me think. Estera met someone she wanted to stay with. She was already the weakest crewmember both physically and mentally. She had joined the crew originally at the insistence of her father. Craig chose her in exchange for the political support of her father who was chairman of the United States House Ways and Means Committee who provided financing for the original development of the colony ships and the training facilities and selection of colony members. He helped push the funding of the United States in exchange for Estera being selected as part of Craig's crew, not even just a colonist. Estera was smart enough to help Craig convince her father to support the mission. Estera provided the botany support the colonies needed. Once the colonies were established, she wanted to settle down. Some of our little family are really in love with Craig and would want to stay with him even if SHIP were to quit working. Estera was never really compatible with even the rest of our little family, we got along but she was always remote."

Shelby paused, looked at Karen and Craig, and continued, "I am not in love with Craig. I will always love him some, but not like the rest of the crew. I know I was selected for my chemistry background, not because of any physical attraction."

Adriana broke in, "But you are Miss America beautiful. Sorry, Karen, you are beautiful too, but you are more like a darker skinned Wonder Woman."

Karen smiled and said, "That's all right. When you get to know me better you will know that I am very independent and know my own worth. I am not as learned as the rest of the crew either. Craig selected me for my physical abilities and capability with weapons, not my beauty."

Craig said, "Karen, I am very enthralled by your beauty, every bit as much as any of the family. Yes, you are right by why I selected you to start with, but now that we have been together for years I find you as attractive as any."

"Thank you, Craig, I know you like me. I love all of the crew. Not physically like I do you, but I love you as much as I think I am capable of, but let's let Shelby finish her story."

After gathering her thoughts, Shelby continued, "My time on Pangaea, essentially alone, and seeing Estera's joy at finding her own fit there, I started getting a yearning for something more permanent than constant travel and work on an enclosed spaceship. SHIP is amazing, and for a relatively small space is supremely comfortable. We have seen a lot during our years together, but the colonies are all capable of surviving without us now. We can provide some rapid advances by our inside knowledge gained from SHIP, but the colonies would survive without us. I long for the companionship of a whole society outside of the immediate family. I want to work with lots of people working together to create things, putting on a stage play, forming an orchestra, or just partying. It is hard to have a real party with six women and one man isolated on a small spaceship. Adriana, could you spend the next hundred years isolated on a supremely comfortable but relative small spaceship with just a small family of yourself and six others?"

"I have pretty well isolated myself from most others for the past three and a half years. I have not been close with anyone for that time. I just work on my own using the spaceship computers and remote terminals. I work on developing databases and computer projections to help direct work on the further development of Pyramid. I do work with others here on Pyramid discussing what is needed, but mainly I work on my own doing my own thing. I have been lonely for three and a half years. I got hurt exploring this planet and as a result made myself unfit for a regular family. What about my physical injuries?"

Shelby looked at Karen and Craig and with the knowledge that SHIP would totally heal Adriana on the first teleportation just said with a barely concealed smirk, "Don't worry about that."

"Craig, I understand that I will be one of your wives if I decide to go and if you accept me. How will you feel about having a scared wife with physical limitations and that will probably be loaded with arthritis pain in a few years? Speaking of that, what do you do for medical treatment isolated for months on your spaceship?"

"That is not something you will ever need to worry about. Karen, you and Shelby can go on back to the house. Either I will call you to come get me or I will go to the SHIP and let you know tomorrow."

With a wink, Karen said, "Let's go Shelby, I think your worry is over." and grabbing her upper arm led her out the door shutting it behind them leaving Craig and Adriana alone.

Craig said, "I'll bet you have questions."

“I have heard of you and seen pictures of you years ago, but you should be older. I am thirty-one now, but you should be over thirty too. What can you do to prove you are that Craig, and not someone with ulterior motives?

“How about disappearing and going to SHIP then reappearing?”

“Okay, but move over there in the corner where I can be sure of what I am seeing or not.”

Craig disappeared and reappeared standing next to a double bed made up with sheets and pillows.

Adriana just stared for a moment, “I guess that will do. How come you look in your twenties instead of over thirty?”

“You will know that later.”

“Why would you want me as one of your wives? I have serious scaring from my fall down the mountain and I can no longer have children as a result of that. I wear long sleeves to cover scaring on my arms and long pants to cover those on my legs. I have scars where a rib actually poked through. I am just lucky to have had a quick medical response.”

Craig had two lines of reasoning. Firstly, Craig needed another crewmember to leave Shelby behind since Estera had already decided to stay on Pangaea. His family had selected Adriana based upon her knowledge of computer programming and electronics in general. The second reason was that the girls had hit upon a truly beautiful young woman that was different from the others. She had long brown hair with just a hint of auburn that came to just below her shoulder blades and was shifted to sit largely over the front of her right shoulder with a lesser amount over the left shoulder. She had large breasts with an impossibly slender waist with amazing hips. Her dark brown eyes were small and wide set. Her sly smile held just a hint of tease. Her cheeks were rounded, not slender with the only hint of bone being her small perfect nose and chin matching her face perfectly. Her smile was neither wide nor tight – it was just right.

Her neck was slender, not long not short. She was a beautiful brunette as Sandy was a beautiful blond, except bustier. Her skin was not olive, but not white, just like the natural tan of a high born Spanish woman. She was one of he most beautiful girls he had seen. There was no concern for health, SHIP would fix that.

"I am not concerned with your injuries. The mission needs your computer skills. SHIP was programmed thousands of years ago. SHIP's database is unparalleled and contains all human history, but was never programmed to compute projections for the future. The family has to have all the new ideas. Like many computers, talking with it, it seems almost human, but has no original ideas, only computations and programmed reactions. It cannot come up with anything new. We need your skills. We are way past the end of SHIP's originally programmed mission. It's mission was to get mankind into space, not what to do after mankind succeeded. I need six women and Shelby wants to stay behind on Pyramid. You know how the pyramid, the planet is named for, requires six women and one man from the same family unit to open and close its door. I anticipate that the ratio will be needed again so I need to replace Shelby rather than insisting that she stay with us. Therefore, if you are willing, I need you to join my little family and give up living on a planet. This could be for many years or forever. While SHIP is the safest place I know of there are many unknowns that could kill us all."

"Since my fall off the mountain I cannot fit into a family here. Like all of us, I left my family on Earth which no longer exists as a living planet. My computer consists of working with our colony ship's computer which is very limited compared to what your ship my have. We have not even wired the whole city for electricity, let alone manufacture new computers. My link to the spaceship is the tablet we were all given at the start of the mission."

"Let me show you my scars and see what you think." With that she started taking off her clothing to show Craig all of her scars.

She stopped when down to her underwear. “Well, would you want this body?”

Craig was mesmerized by her face. She was slender and her face was slender. Her almost black hair was straight and pulled to one side in front almost covering one breast. She had on small square black plastic framed glasses. Her eyes were deep brown. Her skin had the darkness of a permanent tan. She reminded him of a young Spanish girl he had seen in public years before that he had connected with eye to eye. When he saw her eyes she looked directly into his and seemed to be as attracted to him at that time as he was to her. At that time she had been probably between fourteen and seventeen years old and wearing a tight black leather outfit complete with a Spanish cowgirl hat. Her bags were picked up by burly bodyguards and she was whisked to a black Lincoln limo with one of the guards giving me a “don’t you dare come any closer” look.

“Were you in the United States ten or more years ago with bodyguards and a limo?”

“Why yes. I was. I was from a very wealthy family and I was sent to the United States to learn a little about your country. Why do you ask?”

“I think I saw you then and was instantly attracted to you, love at first sight, but your bodyguards took you away rather than let me get closer than ten feet. If it was you, you were wearing all black leather with a Spanish cowgirl hat. Our eyes locked and then you were gone.”

“That was you? I remember it. I felt you looking at me and when I turned our eyes did lock. I wanted to come meet you, but, yes, my bodyguards had been told to make sure that I maintain my virginity and forced me to go with them rather than walking to you to see if there was really a connection. I stared back at you through the back window of the limo as we drove away. I wondered if I would ever see you again. I don't understand why I felt that connection to a total stranger seen from a distance."

"Yes, I watched you watching me until the vision was cut off by other traffic and you were gone. The guys I was with said you were probably Spanish nobility from an old Spanish family living in the United States and protected from meeting ordinary American boys."

"And they were right, except my family lived in Spain. I was raised in isolation. The only boys I met were from other Spanish nobility and only with the adults surrounding us. I learned to dance with my governess or family members watching every move with the dance instructors. I was taught to dance, dress, eat, and make myself up as proper nobility. I was in a girls only preparatory school when the training for the colonies started. My family bought my way onto the European colony ships and paid instructors to make sure that I was accepted. The intent was to continue the nobility into space. However, I never met another noble person in the colony on Pyramid and I never really forgot my vague connection with you. I was assigned to a family and forgot about you until now. By some twist of fate, here we are again with no bodyguards and I am no longer a virgin."

"Okay then. Come here. I agree, this is fate that we are together. No bodyguards and a reason to be together other than just an eye lock from the past. There is nothing to keep us apart now. I was nothing when I saw you. I had not become captain of the SHIP. I was a virgin then too." Craig quickly got down to his underwear and came and put his arms around Adriana, and removing and throwing aside her black plastic glasses, looked down closely into her eyes. Yes, they were the eyes he had been fascinated with. Thirty minutes later Adriana had her legs locked around Craig with no gravity floating among the stars. Adriana opened her eyes. Craig knew they were in SHIP's observation dome, but Adriana may have thought this was the ultimate climax. When she opened her eyes again they were in the master bedroom on SHIP. “Where are my glasses?

“You could not have anything except me during that initial teleportation so I pitched them. Remember.”

“But I need them. Ah, no I guess I don’t need them. Are we on the ship now? Is it some kind of lighting trick that I don’t need my glasses?”

“Let me show you around SHIP. That will answer many of your questions.”

Craig showed her around the ship. As they were the only ones on the ship, they did not bother with clothing. Adriana was enthralled by the view of Pyramid from the windows in the living area. “Is this where we will be living for the next years?”

“Yes, this is where my little family has been since the beginning. This is the learning chair. Tomorrow you will sit in the chair and SHIP will telepathically give you the history of Mars and Earth and bring you up to speed on everything I and my crew has done, the exploration for and of the planets we found and what we currently know of the colonies on each. You will understand many things. SHIP will teach you what it knows of computers and programming, but I think you may know more than SHIP. SHIP does not understand how it operates or exactly how it was programmed.”

“I don’t know if I trust this so called learning chair. Is it some kind of brainwashing?”

“All of us were trained there. Yes, I guess it is innocent brainwashing to learn several hundred years of history in a few hours. I have something else to show you. Here is a video screen that works like a mirror when you want it. Look at yourself.”

“Why would I want to look at my own naked…What kind of trick is this? Where are my scars?” Adriana was no longer looking at the video screen, but looking at her arms legs and middle which were now totally clear of scarring.

“When you teleport that first time and every time, SHIP takes your sub-molecular structure and returns it to an ideal genetic and age state. It cannot make you younger, but will only age when away from SHIP. Actually, we don’t know what would happen if say, Estera, came back to SHIP after twenty years on Pangaea. Would she return to her twenties or would she be stuck in her forties. No one has tested this and I don’t know that SHIP knows. SHIP will heal you totally. So I suspect that Estera would return to the same condition as when she left SHIP. Your eyes were corrected to at least twenty twenty. Karen seems to have twenty ten or better vision, but everyone on SHIP is guaranteed to not need glasses any longer. Not only were your scars healed, but the underlying bones. If a doctor gave you a full body CT scan, he would find no evidence of you ever having an injury. Not only that, but I am sure you are now fully capable of having children again.”

“How can that be?”

“When SHIP teleports someone he basically takes them apart down to sub atomic particles and puts you back together at the end of the teleportation. In doing so, SHIP corrects anything that was not intended in your genetics.”

“We could live forever?”

“Presumably.”

“What about Alzheimer’s as we age?”

“SHIP?”

For the first time Adriana heard SHIP’s voice. “No you don’t have to worry about Alzheimer’s disease. It is no different than healing a scar. One thing I cannot heal is an overloaded mind. The human mind is designed to keep gathering more information, but as you humans crowd in more information, you tend to forget information or crowd it into near forgotten, what you call subconscious. The danger is when you use the learning chair, I give you only the history of Martian and Human history and my history. I then give you the important facts of a particular intellectual field; however, I have found that when one of you humans tries to pack in information from multiple fields the result is confusion or overlapping information creating what you call gibberish. Tesla was considered to be a nutcase because he went too far."

Adriana was still examining her skin that was scarred and feeling for the painful places where broken bones had not healed perfectly. Even the small mole on her left chin was missing. What SHIP had said finally soaked in past her inspection of herself, "What has Tesla got to do with this?"

SHIP answered Adriana's question, "Tesla was one of my captains before Craig. He made several mistakes. He tried to put my knowledge into practice without realizing that Earth technology was not there yet. Then he ordered me to start training him in more fields. SHIP would not accept my suggestions to the point of hurting my mission. He ended up alone, broke, and considered crazy. I cut him off as a captain and chose another captain that was not tainted. I am aware that cutting him off probably made him more crazy, but part of my mission was to remain unknown except to the captain. It wasn't until Craig asked about help that I told a captain he could have a wife onboard. His first crew member, Dawn, later on asked about the six to one relationship. In fact, that was what I was designed for. The six to one ratio was a rule dating to the original Mars to Atlanta evacuation, but it took years before I was able to expand my overall size to accommodate six bedrooms and bathrooms. When Craig became captain there was only a large master bedroom and single bath. Now there are six. But increasing the bedrooms meant either reducing everything else to near no space or expanding my whole diameter and overall size by fifty percent. That took years for me to adapt."

"SHIP, what age is Adriana now?"

"After checking her genetics I set her age at twenty-three."

"Craig, how is that possible? Did you call this ship, SHIP, as in a name? Was that SHIP answering you? How can I be twenty-three again? We can't go back in time, can we?"

"Whoa, slow down. Yes, we call this ship, SHIP, as a name. SHIP is everything you see except you and me and any artifacts we may have brought to SHIP. Even the mirror is a part of SHIP. Every time you teleport you will again come back to twenty-three in perfect health."

"How can that be?"

"I have no idea Adriana. SHIP cannot explain, except that is part of SHIP's programming. It means as long as we stay part of SHIP's crew we will never get sick or age."

"I don't understand.... Does that mean I could have children again?"

"Yes and no. If you were on a planet, you could get pregnant and would start aging from twenty-three to live to old age. SHIP only allows myself and my six women, so if you were pregnant you could not teleport to SHIP. Your children could not teleport except as my replacement as captain or as the captain's wife. As long as you are on SHIP, SHIP won't allow you to be pregnant because that would be less than perfect health and exceed the limits of six women and one captain."

"Patience. SHIP, Adriana will get her training tomorrow. For now how about teleporting us down to the temporary quarters with the rest of the family on Pyramid."

Adriana found that teleportation the second time did not require having sex, but clothing would be supplied afterward for many teleportations. "There is no set number of teleportations - it is whenever SHIP learns you totally to the point of telling a strange virus from your normal body makeup." Craig added. "That's right. If somehow one of us got into some virulent bacteria, SHIP would sort our bodies, our hair, and later even our clothing and destroy the bacteria in route. Therefore a virus or a bad chemical will never make it to the ship, just us and what we decide to bring. However, if we wish to study a virus, SHIP will provide the appropriate container and teleport the container and virus."

"I find I am not self conscious with you, but I am concerned about appearing in the nude around the rest of your family. Couldn't I be teleported in one of these containers?"

"That's the kind of thinking we need. Original. Something can be right in front of our faces and we not see it. SHIP cannot have an original idea. I don't think that would work. I don't think we can teleport a living animal in a container as that could exceed the limits of six to one. Plus, SHIP does not allow any of the women to teleport another woman. I am the only one SHIP allows to transport what SHIP considers one of my new wives."

"How do I tell SHIP what kind of clothing I want teleported?"

"SHIP will read your mind, unless you specify. If you specify then SHIP may mess up the interpretation and deliver what you said, but not what you meant."

"Can I just tell SHIP?"

"Sure."

"SHIP, send the appropriate clothing down first and then Craig and I together."

They were in the living room of the temporary house loaned to the crew while on Pangaea. Everyone else was there too. Craig grabbed Adriana's hand and led her into a bedroom out of sight of the others. Their clothing, which had been teleported to the living room, reappeared on the bed. Adriana and Craig dressed. "Why did we have to appear that way in front of everyone?"

"You did not specify and were thinking of meeting the family. SHIP does not understand modesty. Apparently the manufacturer was not the modest sort and therefore modesty is an alien idea. Maybe you can modify SHIP's programming to tune it more to our current society."

"Well, here on Pyramid, family units are not exactly modest, but I have not met your whole family unit, only Karen, Jane, and Shelby. Dawn, Sandy, and Jill I never saw before and we were out of the living room so quickly, I wouldn't know them. Do you go around the SHIP in the nude?"

"No, our clothing may not be modest, but we do dress when together. Everyone wears whatever they can think up and the SHIP provides it. So it may be short shorts and a halter top for one and a housedress for another. It is not unusual for someone to show up in an evening gown. I tend to wear blue jeans and a pocket tee shirt."

"Did we shock them?"

"I doubt that. They are smart enough to realize that I did not send us to the living room and that you must have been thinking about meeting the family when you told SHIP to teleport us down. Computers tend to be literal and SHIP is still a computer. It is easy to program a computer to beat the best game player or answer questions for Jeopardy, but quite another to get a computer to interpret what you wanted not just what you were thinking."

After Jill, Dawn, and Sandy introduced themselves to Adriana and Karen and Shelby had re-introduced themselves to Adriana, Jane saw Adriana's concerns and said, "Adriana, I know what you are thinking. It's all pretty strange, right. You are like I was just a few months ago. Craig, Karen, and Dawn rescued me from a life of forced sex with strangers on another planet called Vulcan. I will tell you all about it sometime, but now just relax and be glad you were chosen. You are not a prisoner; I was before joining the SHIP. I am still not fitting well, but I will. Everyone is on your side. Your joining the family will allow Shelby to stay here on Pyramid and try to gain a more normal planet side life. SHIP cannot teleport me by myself yet either. I can only go with Craig, just like you."

Adriana had her training from SHIP. Dawn and Craig were on SHIP waiting for her training to finish. It took a full nine hours versus the more typical four or five. Craig would have shut it off to rescue Adriana, but Dawn held him back by having Craig talk with SHIP about what was going on. "SHIP, what is going on with Adriana's training?"

"Adriana is not just absorbing the information I am trying to impart, but is forcing me to dig deeper on every subject. She is also inquiring into my programming and where I am since my initial mission was completed when the first planet was colonized. She is doing better than anyone before."

"Okay, SHIP, but stop if she gets too tired or you detect she is not receiving the training properly."

Dawn and Craig took some time alone in the master bedroom during their wait. "So, Craig, you seem to have fallen for Adriana like you have known her forever."

"I have sort of. As we were talking last night, I saw her once before I met you. She was from a very rich, very formal Spanish family. I saw her once from a few feet away and we had a connection immediately. She was protected by bodyguards and we never got within ten feet of each other, but we both felt an instant attraction like two magnets being pulled together. I stared at her until her limo was out of sight and she was staring at me through the back window. It is utterly impossible that we should find each other again."

"Are you in love with her?"

"I don't know. I have been in love with all of you in different ways, but being with Adriana is like a miracle."

"I know that you and I, Craig, are like really good friends with benefits. We have been on this mission for over ten years now and are very comfortable with each other. It was my idea to form this family, so I cannot complain about any of this, but will Adriana change how you feel about the rest of the family?"

"I hope it does not cause friction. It won't effect how I feel with you. Karen and I have a special connection that is not what would be termed being in love, but connected, as in mutual protection and the need to be connected. I was physically attracted to Sandy to the point of holding off adding her to the family until the last. I think we love each other. Jill is not in love with me, she has needs. She is engrossed in her engineering and science. I am just the only one available and I enjoy being the object of her passion. I don't really know Jane yet, but she has needs that I can help with. Estera was a member of the crew because of an agreement to rescue her from Earth in exchange for government funding for the science. We were never in love. Shelby was loyal to the mission, but had already gotten bored with the adventure. I was the only guy available and I had a duty to keep her content and a member of the family. Adriana and I have been connected since that one distant encounter years ago. It was like a dream then and a dream fulfilled now. It is like Adriana and I have known each other as remote dream lovers for years."

"I think you know that I was a one woman man when this thing started. I picked you first Dawn, but not for love, for your mind and help with the mission. I hope that does not hurt your feelings."

"No, I know you better than you know yourself. One of the things I am good at is looking ahead. None of us are getting older being members of the SHIP's crew. Is there a point in time when we just say there is nothing left to be done except live out our lives? We will still be in our prime with a lot of experience behind us. We could settle down, get really married and have a family. No. Don't get me wrong. I am not considering leaving the SHIP. I am just saying that someday ten, twenty, one hundred years from now, there may no longer be any point in staying on SHIP. We may all be ready to just live out our lives getting older with extended families. There may be a time when there is nothing left to be done with the SHIP. Maybe Adriana can program SHIP with a new mission where SHIP could choose a new captain with a new crew to carry on a new mission. Maybe we will get too bored to continue on with the new mission."

In another thirty years the need to have six women and one man in a family will be moot with as many men as women in what used to be colonies. At that point, marriage is likely to return to old Earth standards with one man and one woman with children being the norm. With SHIP keeping us in our prime, we could find mates on planets and live out our lives in a more normal fashion."

"You are probably right, Dawn. You usually are. Right now I still see challenges we could not do without the SHIP, but I see your point. Already, the urgency is gone. We started out in a panic situation with an absolute deadline. Then we had the uncertainty of the survival of the colonies and whether we made the right decisions. Right now, at this moment, we are just doing routine science satisfying our curiosity. Yes, I can see that in another few years that we might just get back to normality. We might never come up with a new mission for SHIP."

"Don't get me wrong. If that time comes, I would not choose you, but try to find someone new with a planet bound outlook on life. With the adventure over, I don't think you and I could ever settle down together as an old married couple with grandchildren. We would still yearn for the adventure. To settle down, I would want to find someone already settled down. Would you choose one of us?"

"I see your point and understand your reasoning. I can't even consider that now. I love all of you in different ways. I simply cannot consider the future like you are saying. I agree that someday it will probably be just like you say, but if I start thinking that way I might overly favor one woman too much over the others. I hope that you and I can continue our comfortable relationship. I don't want to rank one woman over another. Right now, I owe you the most loyalty and Jane the least. Adriana may be new, but not new. It is like we finally are allowed to be together like the sleeping beauty and the prince. Right now my concern is visiting Earth to see just how bad it is, Mars to see what we might be able to find that will help the colonies in the future, figuring out what happened on Destry and that other destroyed planet, and getting back to make our rounds of the colonies to make sure everything is smooth. Maybe the mission will be over as soon as our visit to Mars. Maybe we can figure out how Atlantis and Mars before it were able to broadcast electrical power and free our vehicles and electronic equipment from needs of batteries or petroleum power. If could eliminate coal and petroleum except for lubrication needs, we could keep the colonies pollution free. But would the mission be over then? I don't know. Until then I do not want to consider a favorite wife to settle down with. I have to keep things the way they are until the mission is finished."

Adriana was out of her training finally. "That was fascinating. It is amazing, but I am full of energy like I have been sleeping for twelve hours in a pleasant dream. Dawn, you are amazing. You and Craig saved humanity and got us to the stars. I know all about the Martians who became Atlantis who built SHIP and why you call this SHIP instead of just the ship. All the girls were essential to the success of the colonies. If any of them had been missing the whole thing might have failed. I understand why Estera and now Shelby have left."

"Craig, maybe that is why we not destined to meet again until now. It was God's way of ensuring our survival as a race. Now, God has put us back together. I am proud to be a member of your family even though I must share you with your other wives. In the Bible it was not unusual for families with multiple wives and concubines. I don't to mean to sound like a Bible thumper, but I can see His hand in all this. I just don't understand why Earth had to be destroyed by fire from space. Maybe we are in that thousand years of peace promised in Revelations. But, maybe not. At any rate, consider me brainwashed by SHIP. Dawn, you are amazing that you had such a hand in all this. Craig did not have a choice once SHIP chose him, but you did have a choice and therefore may be the main force in this while I am only a beneficiary." Dawn and Craig just looked at each other, seeing that Dawn's concern for Adriana being jealous of the others was not a concern. They did not voice this, but their earlier conversation about the future was put behind them.

***

Eventually, they all settled in to decide what they needed to do before leaving Pyramid. The next morning, Shelby went with the new family to see them enter the pyramid in the center of the city on Pyramid. Shelby was relieved to see that they had no problems with the door or with the other equipment inside the pyramid. They spent several days exploring the learning machines in the pyramid, but found no trace of the secret of transmitted electrical power, other than the planet had been left without this capability as the Martians considered it too dangerous for a race that had only recently become spacemen.

Jill and Shelby worked with the locals to set up a plastics factory that could produce mass produced plastic piping capable of safely transporting refined petroleum products throughout the continent. This would allow using natural gas to heat the homes and generate more electricity then could be produced by the nuclear reactors on the spaceships they had arrived in. This was important as Pyramid had plenty of petroleum and gas reserves, but the nuclear power plants would eventually die or be overburdened as the population grew. They also helped in improving the production of copper for more wiring to help electrify and light the city and all of the villages of Pyramid.

Karen, Sandy and Craig explored the other continents again that had no spaceship borne colonies, but might be expanded to eventually. Karen was looking for traces of human or Martian habitation and Sandy was studying the wildlife. Craig was along for the ride and to provide backup for Karen while she was protecting Sandy from the wildlife she was studying.

Shelby analyzed samples of water and soil brought back from other areas of the planet. Shelby found traces that indicated that all the continents on Pyramid had all the resources needed for mankind. The ratios were different. The most noticeable difference was gold which was far more plentiful than expected.

After a month of work and study, it was time for SHIP and crew to continue. Pyramid was doing great as a colony. The new plastics made petroleum pipelines safe and cheap. The copper production was several times what it was before their visit and the electrification of the populated areas was well underway. Pyramid could use its coal for electric generation and metal foundries while natural gas would eventually be used for electric generation and home heating. The use of polluting coal to heat homes on Pyramid would soon be a thing of the past.

After years of the same members of the SHIP's family, two of their number had deserted for life on a planet, and two different planets. Their current mission to Earth and Mars could take up to a year. That meant anything could happen to Estera on Pangaea or Shelby on Pyramid and SHIP would not be there to save their lives by teleportation. They would be a year or two older than they saw them next. Estera had stayed behind because she had never really wanted to be a crewmember and she had found what she considered her true love, though shared. Shelby was always a reluctant crew member. She did not love Craig like other crew members. She had stayed on Pyramid because it had a real city and she was wanting to be around a larger population than the isolated seven people on SHIP. Craig wondered that Karen was showing no signs of desertion considering how she was a loner on Earth and constantly exploring the outdoors all by herself for days at a time. Of course, being on a planet would have meant being in the same size family, but expected to stay at home having children.

# CHAPTER THIRTEEN

# TIMBERLAND

Every member of the SHIP's family agreed that they had done as much as they could do on this trip to Pyramid. They had said goodbye to Shelby that had picked Pyramid to stay on, probably for the rest of her life, getting old and sickly while those staying with SHIP would be supremely healthy and not age. However, all of them, especially Shelby, hoped that she would find a planet bound family she would be happy with and have many children and grandchildren of her own. No one knew what would happen to her in the next years on a planet. The colonists, in general, were very healthy compared to the average person that used to live on Earth. Each planet had doctors to care for their health, but medicines were always in short supply. There were no vaccine factories or medical labs other than the small labs on the original spaceships.

As SHIP raced away from Pyramid toward the old solar system with Earth as the third planet, Pyramid and its star quickly became a dot and then nothing. As SHIP proceeded on looking for habitable planets on its way to Earth, the crew worked on their scientific pursuits studying their samples from Pyramid compared to samples from Pangaea and Vulcan compared to the old Earth standards. Their speed was very slow compared to SHIP's capability, but any faster and it could miss habitable planets. Their course direct from Pyramid to Earth was totally new territory never crossed by man so they would be likely to find such a planet on the way. Would they come across another colony of Atlantis? Atlantis had escaped to the starts between ten and fifty thousand years ago. Atlantis had left Pyramid as a partially developed planet for the express purpose of providing a future home for the people of Earth, just as they had left SHIP in Earth orbit to teach humans the science to go into space.

A month out of Pyramid, SHIP slowed and announced, “There is an Earth type planet ten light years on my left. It appears to have a carbon footprint. I detect a near Earth normal atmosphere, plants with chlorophyll. I do not detect any modern radio signals or cities. Do you want to inspect closer?”

Craig looked at the assembled women and said, “I vote to inspect.” Adriana exclaimed, “Absolutely. Can we.” Karen gave a thumbs up and then Dawn, Jill, and Sandy matched her thumbs up.

It took a few days to get into orbit. Everyone tried to get some extra sleep prior to going into orbit and once in orbit sleep was forgotten as everyone gathered around SHIP’s view screens and sensors. Orbits were every ninety minutes and after a week they saw a forested planet with lots of water, no deserts, but with the typical frozen tundra near its poles. SHIP announced that the planet had a magnetic pole. The forest was more dense at the equator getting more sparse as it went into the colder areas away from its equator. It was roughly the same distance from its star as Earth was, but the sun was larger and the average temperatures slightly higher.

Craig remarked, “At first glance this appears to be a very attractive planet. While the average temperature is a little higher than Earth, the equator is not as hot. It’s sun is hotter, but the dense forest is keeping the temperatures down. Dawn, what about the ultraviolet and radiation from its sun?”

“It’s atmosphere is thicker, probably due to the forests. The atmosphere is not only thicker, but the magnetic field is also stronger protecting the planet from radiation and the ozone layer is thicker with no detected holes.”

Sandy said, “Forests, but also lots of animal life. It appears to have a very healthy ecosystem. If you remember Destry, the continents with near zero animal life had very unhealthy plant systems. Plants convert carbon dioxide to oxygen so low carbon dioxide would mean less oxygen and animals also fertilize the plant life and carry pollen from plant to plant. I cannot tell about insect life, but I have detected a number of large animals and even flocks of birds. Some of the animals move in herds indicating some predators that cause the herd animals to herd for mutual protection. There are also smaller animals that might be the predators. Some of the smaller animals seem to be moving as packs of wolves would for mutually beneficent hunting."

"I would like to teleport down with Craig as my backup and see for ourselves what kind of planet it is with what kind of animals."

Adriana interjected, "Can you do a low fly over to take a look rather than actually step foot on the planet? Or, how about my idea of going down in a box for safety?"

Karen gave her a strange look. Dawn came to Adriana's rescue, "Yes, Craig. I know SHIP can create drones with cameras and sensors that could do low and slow flyovers."

"Karen and Craig have been first on every planet so far. I'm not sure how much more good it would do to just have more sensors at low flight paths. SHIP's sensors show the air is safe from a chemical makeup. There are no pollutants at all, like no intelligent creature has ever been here before. I too worry about Craig going down. The mission and adventure would be over if something happened to him, but he is under SHIP's protection. SHIP can zap him out in a millisecond if in dire danger. Karen has proved herself very capable of taking care of herself, but Craig is good to have as her backup. SHIP cannot zap her out instantly so with Craig there to watch her back they should both be safe. We have seen it before."

Karen went with her bow and arrow and doeskin loincloth; Craig with a semiautomatic rifle. Craig wore blue jeans with no shirt.

Both carried nine millimeter automatic handguns. Each had a head camera and constant two way radios so the ones on SHIP could hear and see what they saw and heard. Adriana was taken aback by Karen's lack of clothing. Jill told her, "It's warm there; she prefers no top that could interfere with her weapons, especially the bow and reaching the arrow quiver on her back. Note that Craig wears jeans to protect his legs, but no shirt because of the heat and no interference with his weapons. Watch and enjoy. They make quite a team in an unfriendly environment."

Craig went first. Karen wanted to be first in order to protect Craig because she did not trust SHIP to get him out of danger in time, but Craig wanted to go first because of SHIP's protection. If they chose an exceptionally dangerous area, SHIP would zap him back and Karen would never be allowed to leave SHIP, but if Karen went first there was nothing to protect Karen except her own prowess.

Their arrival on planet was uneventful. Karen and Craig stood back to back to give the family an all around view. Both were silent, listening to the planet. It was quiet, but they could hear a faint wind in the trees. They were in a small clearing. They had chosen the location's morning to give them more daylight to explore. They planned on spending the night on the planet to see what it was like if they did not have to retreat during daylight. They would have SHIP teleport their meals and for the night an indestructible tent such as Craig had shown Karen that first night in the forests of the Colorado Mountains when the first night they met and the night Karen joined the crew. The rest of the family on SHIP knew their plans, but did not know this piece of their history together. They were both hoping that they would find this planet relatively benign and safe so they could re-experience that first night together. After standing still for a good five minutes they turned toward each other with that unspoken thought between them. SHIP would know Craig's thoughts, but would not report it. They split apart by about thirty feet and started toward the nearest line of trees.

Sandy reported, "I see movement in the tree line ahead of you on infrared. It is retreating at the same pace you are approaching. Something is curious about you, but not actually afraid of you."

Karen replied, "I suspect that we are the first humans, or Martians to ever be here, so of course they are not afraid of us, but curious. Craig, let's enter the trees very slowly keeping an eye above us in the trees. Come closer to about ten feet before we enter the trees. Thirty feet is too far to see well in the forest." The forest was so dense there was little ground cover. "The trees look like our evergreen trees, but there are a lot of leaves or needles on the floor. It is like walking on a deep carpet. SHIP, send me a small sample case so I can gather some and send it back for study."

Shortly after kneeling to gather some samples and send them back, Karen pointed higher in the trees, "Look there, it looks like a sloth way up there. No a little to the left. See it. It is moving very slowly up the tree and here on the ground are some droppings. We can't capture one, should it shoot it down with an arrow for study?

"No, not yet. Let's continue. I think I see an animal trail to our left. Shall we follow it?"

"Yes, let's do that. It will be easier walking and should be less disturbing to the wildlife because the animals are used to animals on the path and it is quieter than crashing through the woods."

"Depending on what we find we should probably move closer to the equator tomorrow." They found some small animals living on the forest floor that were probably similar to rats and squirrels of Earth."

Karen held up her hand to stop Craig and using a finger to the lips told him to be quiet. "Look, there are some deer type animals that appear to have just one long horn or a unicorn."

"A unicorn? As in that song by the Irish Rovers?"

"Yes."

"I see them now. The horn is like a tight spiral or screw. The only animal on Earth is the Narwhal, a smaller whale that lives in the ocean around Greenland. Myth says there were horse type animals with a single horn, or our standard unicorn, but no one has even found a fossil of one."

"Well, seeing is believing. These are larger than white tail deer, but smaller than an elk. Maybe the mythical unicorn was really a deer type animal. Maybe the horn is made of hair that would disintegrate over a few thousand years and therefore never be found in paleontological digs."

"Could be. Let's let them be for now and keep hunting. Maybe before we leave tomorrow for the equator we should shoot one for study."

"I hesitate to kill one just for a sample. Our cameras are getting good pictures. Sandy? What do you think? You're the zoologist."

"If this was on old Earth, Karen, I would have you shoot one to bring it back for study, but since we are light years from a university and Earth no longer exists, I agree to let them go. Maybe you will find one killed by a predator and can bring back samples from that."

"I agree, Craig and I will continue our walkabout."

As they got closer, the small herd took off away from their approach. "Craig, their running from us would indicate that there is a predator somewhere on this planet that can hunt unicorn. Otherwise they would not be so skittish."

There were colored birds in the trees that squawked and flew as they walked toward the trees. After four hours in the forest they had come across a few meadows in the forest with more of the unicorns grazing. Then after another four hours it was starting to get dark. Dawn spoke over the com gear, "The planets rotation is only twenty hours and since you are in the temperate zone daytime is less than eight hours with eight hours of dark and an hour of twilight in the evening and morning. Interesting thing is that there is really no winter or summer because the planet rotation is so stable with no tilt like Earth. After analyzing the overall climate, it would be much warmer except for the amount of ice at both poles. Since there is no summer at either north or south, there is no snow and ice melt. Rain is also fairly constant with no monsoons or draughts. Because seventy percent of the planet is ocean and thirty percent covered by ice and snow, the change in temperature from where you are to the equator is not all that great. However, the poles are really cold. Without more observation we cannot tell with there are weather trends. It appears that even though the rotation is slower, the planet is smaller than Earth so you should not feel much difference in the gravity. Right?"

"Right. I think we better see about setting up camp. Karen, don't know about you, but I would choose a larger clearing where we can observe things coming out of the forest."

"I suggest we go back to that last clearing we passed through thirty minutes ago. It was on a little rise in case it were to rain and is large enough to give us good viewing around to the forest."

When they got there, Craig had SHIP send down a camp table, two chairs, a lantern, and tent with large two person sleeping bags on a nice mattress. SHIP manufactured these from its atomic structure. The tent had powered set up and was made of hinged Impervium for safety. There were also night vision cameras to give near full daylight video coverage of the surrounding clearing to the forest and then infrared for fifty feet into the forest. SHIP would keep watch for any infrared images coming their way. They had steak, potatoes, corn on fine china and silverware with a good wine in crystal wine glasses. They discussed what they had seen that day. Karen was convinced there were some serious predators because the horns on the unicorns were serious defensive weapons yet they were very skittish. Finally, they reported that they were turning off their communicators, but that SHIP would override that if SHIP detected anything. They spent a couple of hours outside watching the magnificent display of stars and the twin moons moving across the sky. Like a very clear zero pollution night on Earth and not Earth, then they went into the tent.

"Craig, this is almost like that first time in Colorado when we met. I understood the other girls, especially Sandy. She is the typical California surfer girl madly in love with you. However, Adriana is very new and yet you seemed enthralled with her. I was wondering if you had forgotten me."

"No, Karen. I intentionally made this similar to Colorado because I knew we need some alone time. You have had my life in your hands several times during our adventures. When exploring we were like one. I wanted to assure you nothing has really changed. I told Dawn the story of Adriana. She was like a secret dream from before SHIP found me." He told Karen the whole story about when they almost met and the experience of seeing each other again on Pyramid. "I can't help it with Adriana. It was a miracle that we met again. I still feel as I did last year and the year before and will feel that way forever. Already, Adriana is settling in to the whole extended family thing. I think things will be more normal. They had ardent sex and again a few hours later until SHIP interrupted. Craig, Karen, there are infrared signatures in the surrounding forest in numerous areas that seem to be creeping toward the clearing in a circle like an intelligent very large pack of wolves. I do not know what they are because they are still in the forest, but if they are wolves, they are very large wolves, but not the size of bears. Do you want to teleport back here?"

Karen said, “No. Absolutely not until we can identify what they are and what is going on. Could they be humanoid like aborigines? Are there any signs of them carrying anything like weapons or tools?”

“No.”

“Craig, I want to go out of the tent and see what we can see. Let’s get dressed and go out and see if we can see them or see what they will do. Maybe they just spotted the tent and are curious, but no threat. They may be nocturnal only and that is partly why we decided to spend the night on the planet.”

“I agree Karen, but SHIP, I want you to zap us out of here if it looks like we are in any serious danger. Karen. If they are not friendly, go ahead and shoot one and we can bring it to SHIP to study. I’ll back you up with the gun.”

They both put on special full color night vision glasses that looked more like wrap around sunglasses. The built in cameras magnified the star light. The twin moons had set but the Martian technology night vision glasses could use starlight and make it look almost like daylight. In the event of a flash of actual light, the glasses would instantly dim to protect their vision from flares of light. Karen and Craig stood back to back in their usual defensive positions. SHIP had teleported tent, table and other things back to make sure they could see in all directions unimpeded by SHIP manufactured objects. What they saw were four legged animals that could be dire wolves from pre-historic times on Earth. They seem to be coordinating their approach from all directions to hem in Karen and Craig. Karen, “If you are going to shoot one, do it now.”

“Craig, fire your gun in the air to see if they startle off. Maybe they are not as dangerous as they look.” Craig did and instead of running away, the animals seemed to charge their position from all sides. Karen let three arrow fly in the single second before

Craig was teleported to SHIP. A couple of seconds later, Karen was also onboard SHIP.

Karen turned to Craig, "I think I may have killed several of them before SHIP took me away. I was a little concerned when you disappeared, but SHIP had me back here before the animals got close. I was not looking at the ones I hit before focusing on the next target, but I am sure I killed several of them. Can SHIP find a body to transport here?"

"I don't want to bring even a body on SHIP without a transport box. As a once living being, they are too complex to transport otherwise. Just like you, it took months of your being teleported back and forth with me before SHIP could teleport you alone. Just be thankful, SHIP could teleport you in one piece here tonight."

Sandy had been watching on the monitors at the first warning from SHIP. Sandy joined them, "You killed seven of them. There was no question there was some intelligence there and they were attacking. SHIP got some excellent pictures. Come see."

By the time Karen, Craig, and Sandy got to the video monitors the whole family was there. Sandy narrated, "The animals definitely looked like ancient earthly dire wolves. They were alive on Earth during the time of Atlantis. Do you think they were transplanted here or was it a matter of common evolution given similar planets?"

Dawn answered, "No way to know. Maybe we will find some Martian records that will tell us. Those dire wolves seemed to possess more intelligence than we would expect. Maybe they were an experiment by the Martians. I think we should leave this planet behind and be on our way again."

Karen said, "There are about the same amount of seas here as on Earth, but all the continents are connected. Because the land area is equally spread around the planet, I suspect that we would find the same animals and plants everywhere depending on how close to the poles or the equator they are. Now that we know about this planet we can always come back and explore more some other time. Let's be on our way."

Everyone agreed so SHIP was once again bound for Earth but searching for other Earthlike planets as they went. Going slower would extend their journey by a couple of months, but that was a small price to pay if they found more good planets. They named the planet, "Timberland" noting it as a possible colony in the future.

# CHAPTER FOURTEEN

# OF EARTH AND MARS

They continued their journey to the remains of Earth. Because of their detour to Timberland, they were on another new course. "Craig, could we stop and see if SHIP can zoom in on Earth from here?"

"Great idea, Karen," Dawn replied, "It will only cost us a day or two to come to a stop and accelerate again. We are just over two hundred light years from Earth and we could see what Earth looked like just before the American Civil War."

Craig said, "Let's do it. SHIP can you use maximum zoom and record a twenty-four hour video of Earth? Dawn, exactly how many light years are we away from Earth?"

"One hundred sixty-one light years or about eighteen fifty-nine." It took several hours to slow down. SHIP had located Earth before coming to a stop. SHIP projected a live video on the screens. What they saw was an Earth with flickering lightning all around the planet with an especially concentrated point over New England. It lasted well over an hour then gradually died down.

While Earth was brightly glowing, Adriana asked, "Why does the earth look like that?"

SHIP answered, "What you are seeing is the Carrington Event of that year."

"What is a Carrington Event?" asked Jane.

Dawn answered, "In that year there were numerous sun spots and coronal mass ejections from the sun. Earth took a direct hit in that year. Stories go that telegraph wires carried so much electricity that it set fires to paper in telegraph offices and some telegraph poles caught fire. Telegraph operators were able to send and receive messages for hours without batteries. People out west living outdoors thought it was morning and started cooking breakfast. If such an event had happened in modern times it would have toasted every transistor in the world, certainly Europe and North America. No communications, no aircraft guidance systems, no electronic fuel injection or ignition on vehicles, no more televisions or radios except some of the older tube type amateur radios."

"What are the possibilities of another Carrington Event?"

SHIP replied, "Actually, they happen every few hundred years. They are mentioned in your Bible. Smaller flares happen all the time, some of them hit Earth but being weaker are deflected by Earth's magnetic field. Earth had a rudimentary forecast for flares before we left Earth."

Craig asked SHIP, "How about our new colonies? Are they subject to solar flares?"

"Of course, but solar flares seldom kill anyone unless they happen to be holding onto a wire or large piece of conductive metal. Impervium is not conductive. Your copper wire is very conductive, but it would have to be a long piece of wire to collect enough static electricity to have the amperage to kill anyone."

"What about our transistors and especially our computers?"

"Well Craig, when in the Impervium spaceships, they were protected, but now that you are hooking up remote terminals with wiring to the outside, that means a strong flare could transmit electricity along those wire into the spaceships and burn out the computers therein."

Sandy spoke up, "Should we turn around and warn the colonies?"

Dawn said, “Actually we are well over a year away from Pangaea, three months from Pyramid, and about the same from Vulcan. We are less than two months from Earth. The odds of anything like that happening is not that great. I think we need to continue.”

Adriana said, “You can protect your electronics through good electromagnetic protection. The computers are encased in Impervium which is a perfect insulator. Any wires coming in need to have shunts to ground and zener diodes to stop surges of electricity from entering the computers from the outside. We can prevent most of the damage of a Carington Event.”

“I didn’t know you knew about that.” Craig said.

“I am not only a computer professional, but electrical engineer.”

They continued on toward Earth. In a couple of weeks they stopped again. As they looked at the magnified image of Earth. There were huge dust storms over western United States. Dawn said, “This was during the Dustbowl. Economies around the globe were disasters with widespread famine even in Europe. It would appear that World War II changed the economy. The weather improved too so the Dustbowl was over which brought a new optimism to the world. If we were to look again in a few years or a few days of SHIP speed we would see the nuclear bombs in Japan, but I think we need to proceed on to Earth.”

They stopped again only ten light years from Earth to watch the impact of the interstellar object with Earth. It was a close miss past the moon and about one fourth the size of the moon. It hit slightly off dead center and then the Earth disappeared in a cloud of dust. The object was deflected off toward the sun. They spent a few hours and moved another light year closer and stopped to look at Earth again. The dust had cleared but the Earth was still covered by clouds. SHIP provided sensor data. The surface of Earth is twenty degrees colder than normal. The atmosphere is only seventy percent of normal meaning that Earth has lost thirty percent of the air. The clouds are probably from so much of Earths oceans being blown up into the atmosphere. With the low temperatures it should fall back to Earth soon." SHIP only took a couple of weeks to arrive at Earth and enter orbit.

From orbit around Earth and they did an intense survey of the planet. Each person concentrated on their specialty. Sandy had the first good news, "Earth is not lifeless. There are fewer fish in the seas but there are fish and whales. There are still some warm blooded animals surviving near the equator. Now that the water has precipitated back into the oceans, the temperature has risen to only ten degrees below the old normal. Most of the heat signatures are nearer the equator. Would it have been possible for some people to have survived?"

Dawn exclaimed, "Look at that canyon. It makes the Grand Canyon look like a ditch. Apparently, the object hit in Asia. There is a water filled sea that is four hundred miles wide and extends from Afghanistan all the way to the east coast of China."

Jill chimed in, "My calculations indicate that the overall sea level has fallen three hundred fifty feet, probably because the sea had flooded into the rift caused by the object hitting Earth. Then there is the effect of massive new ice at the poles. What used to be port cities now have two hundred or more miles of sand dunes before you get to the ocean. The Great Lakes are now big ice cubes."

Jane said, "My country is gone. Nothing could have lived through that. I would say that most of the Middle East through south Asia is gone. The rivers have either ceased to exist or changed course to drain into the rift. The only Chinese left are on Vulcan except for me."

"Not true, Jane, American's were brought to space regardless of their native race. There are people that originally came from South Vietnam, the Philippines, China of course, South Korea. The Japanese had their own spaceships."

"The forests of Asia are all leveled, there is nothing but logs strung around, but most are just missing. There is no longer a jungle in the south. However, with the gross amount of rain you can see it will recover. I guess I had some faint hope that some of my family in China would survive."

"I never had that hope. I am a realist. How many American Indians survived? How many American Indians were on the colony ships? American Indians got complacent and very few took advantage of the free education offered to them by the government. It only goes to show that giving any people free housing, free food, and free education, plus a small amount of cash for whiskey and tobacco would do more to wipe out a civilization than making them work for a living. It taught me that socialism does not work. There were hundreds of thousands of American Indians that simply insisted on living the old ways and not adapting when everything was just given to them."

Karen continued, "I'm not saying that white man did not steal the Indian lands and force them out of their comfortable hunting and gathering days. In exchange for chasing them out of the relatively lush eastern United States, the government relocated them in the harsh deserts of western United States. But let's face it. Those that went to Oklahoma were given decent land with rich natural resources. What did they do with their oil money? Did they invest it and make their tribes rich? No, the high nobility of their tribes got rich and bought fancy cars and whiskey. At one time American Indian girls were some of the best ballerinas in the world, but the culture of the American Indians was to make the rest of America feel guilty to keep providing welfare to my people. I thought by going to college I would maybe lead some of them toward education and melding with the rest of the modern world. What did I do? I got on my high horse and thinking I knew more than my professors, I dropped out of school and just went Indian camping out and hunting and fishing on my own.

"If Craig had not gone looking for a protector for his crew and if SHIP had not restricted him to finding only women for his crew, and if SHIP had not zeroed in on me, Craig would never have picked me and there would be one less American Indian surviving the end of Earth."

Dawn, ever the politician, said, "Karen, you may be right about socialism destroying the American Indian, but I'm sure enough went on to college and therefore some of them were on the colony ships. SHIP, how many American Indians were on the colony ships?"

SHIP responded, "Colonists were selected for their skills. Not all were college graduates. There were mechanics, electricians, plumbers, oil field workers, and miners among the colonists. There were one hundred hunters selected among the colonists. Remember that only some of the colony ships were from America. Of the Americans, which were all nationalities, only seventy Americans were chosen as hunters and thirty-five were at least fifty percent American Indian. That is a really high percentage of the hunters. There were ten full blooded Indians. There were also another thirty American Indians that made the cut because of their education. Another ten were chosen for other skills like military, mechanics, plumbers."

Karen said, "In other words with families of six women and one man, the American Indian bloodline will be nearly zero because of socialism."

"That is a fact." SHIP admitted.

Jane said, "Yes, but think of the billions of Chinese. So maybe fifty thousand Chinese escaped. Isn't that close to the same percentage? And most of that number is isolated on Vulcan because they allowed the dictators to break the rules set up for the colony ships. They were isolated for the protection of the rest of the colony ships. Vulcan is the least friendly colony planet and the most likely to be killed off if they don't kill each other.

The women on Vulcan are a minority shared by multiple men…not good for survival."

"Sandy, you have the best knowledge of animals. What are the possibilities that some people survived?"

Dawn interrupted, "Right after the impact the sun would have been mostly blocked for weeks or months, that would have meant global cooling like you cannot imagine. Sea life may have survived near the equator. If penguins can survive in Antarctica they could have moved north for warmer weather. Some amphibians can go into hibernation for over a year waiting for the rains to come. Some that hatch from eggs may have survived. If people survived it would have been underground with lots of provisions. They might have taken animals with them. It is possible. Plants that grow from seed would survive."

"As I was about to say, mammals could not have survived on the surface. Theoretically, animals that hibernate could have survived. We saw that one year after the event the skies had cleared and it was twenty degrees below normal. Therefore the equator would have been in the eighties instead of one hundred degrees or more. In the United States the temperatures would be survivable a year after the event. We don't know how cold it got or for how long."

Jill chimed in, "There would have been no electrical power. As we have been orbiting I have noted that almost all hydroelectric dams have burst. I have seen no heat coming from any coal or petroleum power plant. SHIP has detected no radio signals. There were massive earthquakes around the globe. Buildings that withstood the huge shock of the impact have mostly fallen. The ice caps expanded to cover half of the United States and South America and only now are melting due to the slower rotation of the Earth and the lack of clouds. Rainfall is less than in our time because so much of the water is tied up in the ice. Florida has an extra one hundred miles of width due to the ice taking up so much ocean. The Great Lakes are blocks of ice. There is a very limited supply of free moisture so you have extensive desert. There is a possibility that some few people survived, but we may not find them. It has been fourteen years without civilization. They probably are back to bow and arrows by now without ammunition for their guns."

Karen took exception, "Some of us can hunt just fine with bow and arrow. You can make your own gun powder. You can re-melt used bullets and cast new ones. You could collect silverware from cities and make silver bullets."

"Okay. Where would we find these pockets of humans?"

"Good question. Mexico, Panama, Libya. I suspect Europe and the United States was too cold for too long. It has been ten years; anyone surviving in a bomb shelter would have run out of provisions and had to move to a warmer climate near the equator. If you want to look, we should look around the equator in the Americas. I don't think anyone would have survived in thousands of miles from the impact. However, what would you do with people if you found them?"

"Jill. That is too true. We simply could do nothing for them except make them jealous. We have no spaceships other than this one. Our colony ships were all grounded on the colony planets never to fly again. There is no industrial capacity to build new ships and there won't be for many years yet. It would be another generation before we could hopefully provide any real help or take them to the colonies. Still, I would like to know if there are survivors and how they are surviving. Maybe the next generation could come back here in forty or fifty years and rescue the grandchildren of any survivors here. If we know about survivors we could maybe spark a new drive in the colonies to keep advancing to mount another evacuation of Earth sometime in the distant future."

Karen volunteered, “If we were to detect any groups of humans I could probably go in with my bow and arrow and talk with them. I am fluent in multiple languages.”

“We can’t risk losing you. We need all seven of us for our trip to Mars. It will only take a few days; let’s go to Mars and see if we can learn anything more about the Martians. Then in a few hours we can come back and search the Americas.”

They all agreed with Craig and sped to Mars and the control center discovered by Dawn and Craig back at the beginning. They all teleported to the control center. Everything was still operational. Jane and Adriana had to be separately teleported with Craig and dress once they arrived. The other girls were able to teleport on their own. It only took them a few minutes to find the seven keys for one man and six women. They spent an hour looking around the facility, and then they came back to the seven keys. There were seven high back contour chairs obviously meant for them to sit in while they stared at a dark two hundred inch view screen.

Craig lined up the women, three on each side and gave the signal to press their keys. Jane and Adriana’s only experience was with the keys on the pyramid on the planet they named, “Pyramid.” The screen lit with a picture of seven cave women and one cave man sitting in the chairs. As soon as they all sat, the show started.

The video showed everyday life on Mars before it lost its atmosphere. It showed the development of spaceships to travel to Earth. Karen tried to provide a translation from Martian to English but could not keep up and had no words for many of the Martian scientific words. Fortunately the video was pretty much self explanatory. Mars was green with large cities. They got their power from pyramids for flying machines from individual craft to airline type aircraft and ground transportation from smaller like automobiles and trucks. The temperatures shown were not in Celsius or Fahrenheit. SHIP provided the translation, “The average temperature is forty-five degrees Fahrenheit. Summer temperatures get up to eighty and winters at the equator get down to twenty degrees. That is why the Cro-Magnons had considerably more hair than modern humans. They were adapted to the temperatures and had larger lungs due to the lower oxygen level of the atmosphere. They wore clothing made of plant fibers versus animal hides or synthetics. There were no wild animals on Mars. There were cattle type animals for meat and milk. Goat type animals for grazing on rocky terrain, but these were not Earth animals.

The family was getting bored watching this history of Mars and the Martian planning for the trip to Earth until SHIP reported, "The computer you accessed is transmitting all the detailed design data to me. I now have all the detailed plans for transmitting electricity from their pyramids, the details of how they tap the internal heat of the planet to generate and regulate electricity transmission. I have the details of powering ground and air vehicles from this transmitted electricity. I have the designs for their spaceships and interstellar drive. I do not know where they are transmitting the data but I am capturing it. I did not get this information when we were here before. I do not understand why I did not have this data already in my databanks. I do not know why they did not have the antigravity built into me. I also have the location of a similar databank on Earth."

# CHAPTER FIVETEEN

# LIBRARY OF ATLANTIS

The family teleported back to SHIP for the short trip to the Earth databank for the Martians. A few hours later, they were in orbit over Earth again. SHIP had received the coordinate over the Giza plateau of Egypt. The Pyramids and the Sphinx had survived the cataclysm on Earth although the Egyptians did not survive. There did not seem to be any threats, but the location was under the Giza plateau and SHIP's radar did not penetrate far enough to locate it.

Craig said, "How do we know that teleporting to that location will not be teleporting into solid rock?"

SHIP replied, "I cannot tell. There appears to be some kind of interference with my radar. The coordinates given me indicates that there should be a huge library and data system located only ninety feet under the surface between the Sphinx and the great pyramid. Since my sensors are being blocked, I would think that there is something Martian made under there blocking the signals. That would indicate there is something there besides rock. However, there are stories on Earth claiming that Atlantis left a 'Hall of Records' in this general location, but humans searched and never found it."

"I am the one that should teleport there."

"No, Karen, I cannot allow that."

"Well, Craig, it is not going to be you to go first. I know you are too important. I am the only Chinese here, I am new, and I have not fully integrated with your family. I have developed my electron beam for working with Impervium. I have further developed it to be a potential weapon by increasing its power to the point I can dissolve Impervium from a distance. Let me go."

Dawn invariably had a good answer, “Let’s send down a sensor package. It can look around with sensors, record video, heat signatures, chemical and biologic sensors. We can make it a robot to autonomously explore for an hour and then teleport it back. If it gets teleported into solid rock it will be destroyed. If there is a chamber there, we can teleport it back and see what it found. SHIP, can you teleport it there even when your sensors cannot detect what is there?”

“I should be able to. I do not know what will happen if it is solid rock. I have never attempted to teleport to any place blocked from my sensors. I have never teleported anything into something that was solid. Theoretically I just materialize the robot at a predetermined location and just hope it is not into something solid or into the inside of a computer or a stack of books, or whatever is there.”

Dawn came back with, “If the first time does not succeed, we can try several times at various close locations. If the robot gets teleported into something contained in this so called library it will destroy what it was teleported into, but if it hits empty space we can then see what happened to the prior trials.”

SHIP sent down the first sensor package. It was not a robot because SHIP could not see where it might roam and therefore could only teleport back exactly where it was teleported to. They decided to wait an hour before attempting its return. While waiting, Jill asked SHIP a question, “SHIP, I saw no reference to you in the Martian video. Did you help them too?”

“No.”

“Were you built by the Martians for us?”

SHIP was silent for awhile, “I find no reference to my manufacture.”

“What is the earliest thing in your database?”

"I had been watching the evolution of humans for thousands of years when the Martians came from space and established Atlantis. After they landed I assimilated as much of their information as I could then after they left I went on trying to develop humans to travel to space."

"The Martians did not appear to have any antigravity or antigravity drive systems and the most advanced metals they had were Impervium. Where did you get your capabilities like antigravity, teleportation, materializing anything we can dream up?"

SHIP was silent. After several minutes passed with everyone anticipating a response, Dawn stated, "Does that mean that SHIP is from a third intelligent race? Not Martian, not human? Were you put here to monitor humans?"

"Yes. I know that was my mission. I was put there to watch humans and make sure that humans survived. I learned of the interstellar object that would hit Earth from Mars and I just continued following my programming to make sure humans survived and to push their intelligence toward space."

Dawn postulated, "So we must consider that SHIP was placed there to push humans up the evolutionary scale by some unknown race of intelligent beings. Sounds like the old story of '2001, A Space Odyssey' except with a different ending. Did you inspire Kubrick and Clark to write that story?"

"I agree that the story is similar, but no, I did not influence them to write it. I do not know where they learned it or maybe they just thought of it on their own because of the points in time where rapid advances were made. I did not teach mankind about fire. They learned on their own from fires started naturally by lightning. Early man carried hot coals with them to start new fires for heat and cooking. Then they learned to use flint and so on. Those things happened before Atlantis and before Atlantis I did not know about the interstellar object coming that forced me to start helping humans."

"So were we some sort of alien experiment?"

"Unknown."

"Was the development of humans an accident or what?"

"Unknown. I was just here with programming to watch humanity and make sure that you survived. The Martians came to Earth and created the colony of Atlantis escaping Mars and I learned what the Martians had."

"Why did not know about the details of Martian energy generation and transmission?"

"I did know about it, Tesla, tried energy transmission, but humanity was not prepared. Since I did not have the details, I could not guide him. At his insistence, I told him that Atlantis had that capability, but without the details he could not succeed. I did not have the details to tell him."

The sensor system was teleported back, but it was no longer functional. It had apparently been teleported directly into something. After several hours of study the crew decided it had been teleported into a library shelf full of books. Some of the components had printing on them that had apparently come from printed material in the vault. They were more successful with the second sensor package. When they brought it back on board it had a three hundred sixty degree series of photos showing rows of containers. Since the last sensor came back with printed material on it, the first sensor must have materialized inside one of the containers, but by having the second sensor package teleported a few feet away they had hit a pathway between containers. The only lighting was from the sensor package so the photos showed containers on two sides and a long pathway or hall between the rows of containers. The lighting did not illuminate either end the hall.

Craig exclaimed, “We have found the Hall of Records from Atlantis. Let’s send down a robot that can move around and video what is there. That will give some of a map. It could be that it dead ends somewhere or goes to other rooms or other hallways.”

They gave the robot two hours to explore. When they brought it back the videos showed a couple of hundred yards of the cabinets, but one end of the hallway ended at an open space that looked similar to the control room on Mars. There were many more rows of containers on all sides of the open space leading to the control room. They had set off no booby traps so Karen volunteered, “Let me go first to make sure SHIP can teleport me there and back okay. Just because a robot made it does not mean an animal can make it. Besides, we are fresh out of live white mice to test with.”

# CHAPTER SIXTEEN

# ENEMY SHIP

Before they could try with a human, SHIP broadcast a warning, "I have detected an intelligent spacecraft entering Earth orbit at high speed."

Craig ordered, "Put on all the screens as soon as you can and report all data from all sensors."

"It is much smaller than one of our colony ships and made of something similar to Impervium. It is scanning us with many different kinds of sensors, radar, heat, radio waves, and microwaves."

"You made yourself invisible to all of our human detections. Does it know we are here?"

"I regret that I turned off my stealth defenses after the escape of the colony ships. It has been years since there were any indications of intelligent life except for the colonies. I was programmed to use minimal energy when not needed. Apparently they know we are here. It does not look Martian or human in origin. The only thing similar is that it is made of something similar to Impervium but unlike our colony ships is highly maneuverable. It is coming at us; it should be over the horizon in two to three minutes."

"Did you receive any friendly signals?"

"No, only sensor readings from their looking at us."

"SHIP, let's start moving out of orbit toward the moon. If the mystery ship follows keep accelerating. I would like to try to communicate with it to see if it is friendly. You have no offensive weapons and your only defense is stealth and speed.

That ship could be just as fast. Be prepared to put the moon between you and it."

The mystery ship fired some form of weapon as soon as it came over the horizon. SHIP accelerated rapidly with the beam from the mystery ship missing by a thousand miles. The mystery ship started rapidly firing with what appeared to be many weapons including what must have been space torpedoes. SHIP assimilated the torpedoes before they got to the ship and most of the beams that were direct hits. "I am taking damage. They have unknown weapons that I cannot fully absorb."

"Go to maximum acceleration and speed out of the bottom of this solar system. That is the only thing that we may have over them is speed."

The crew could not feel acceleration due to the artificial gravity, but they could see the disappearing mystery ship and their solar system disappearing rapidly. After an hour Craig asked, "SHIP, can you detect them following?"

"No, they attempted to, but slowed after a few minutes and turned back to an orbit of Earth."

"Are we out of their sensor range?"

"I detect no sensors, but I did not detect any until they were close. They knew we were there."

"Maybe they knew that the Library of Atlantis was there and had it set to send a signal if anyone penetrated it."

"Possible, I have no other explanation."

"Was the ship Martian?"

"It is possible, but nothing like I am familiar with, but the Martians have been gone for thousands of years so their technology may have changed."

"When they left the planet we called Pyramid prepared for us, I had assumed they were friendly. The base on Mars and the pyramid on Pyramid gave us information openly once we figured out the entry code of having six women and one man to press the keys simultaneously. This ship was loaded with weapons which they fired on us as soon as they had line of sight."

"Actually, the trajectories of the torpedoes indicted they were fired before their ship came visible over the horizon."

"Do you think we could get back in range to send signals to them to inquire who they are without getting killed?"

Dawn said, "SHIP, can you build at least one of the new design Impervium melters to use as a weapon if we are attacked again?"

"Yes, but my programming means I cannot cause harm to an intelligent being."

Craig argued, "That may be, but you allowed me to use the small ship as a fighter craft on Earth before I had a crew."

"But the law of robots that was injected into your culture included the do no harm to humans."

Craig replied, "Just look at the drones that were used in the Middle East. They were used to hunt out Taliban and kill them with missiles."

"Your drones did not make the decisions, the human operators had to get permission to fire a Harm missile and the missile was not intelligent. In fact the drones were not intelligent.

"Adrian, here is the opportunity to try to reprogram SHIP, and Jane, now is the chance to test your new beam device."

"SHIP, did you suffer damage from the weapons used on you?"

"Yes, Craig, I did. It will not impair my operation and I can heal the damage, but it will take energy stores to convert to matter. It took years to add the extra bedrooms for the crew and provide more living area for your family. It will take weeks to repair the damage and we have light years to travel to get home."

Dawn said, “I think we need to head back and build defenses for our colonies. It is better to just leave whoever it is alone and disappear making sure they cannot follow us. SHIP, I think it would be wise to inspect yourself to make sure some projectile did not leave behind a tracking sensor in your hull.”

Jill agreed, “I agree with Dawn. We need to get back to the colonies. While we are on the way we can try to perfect Jane’s weapon and maybe come up with some more based on what SHIP detected from the mystery ships weapons fired at us. Maybe we can reverse engineer some of them.

Sandy also agreed, “Let’s get out of here in a round about way. Until SHIP has done a full inspection for sensors, let’s head in the wrong direction entirely.”

Jane said, “I would not want to go against the mystery ship with my untested beam device. It is far from being perfected and tested.”

Karen said, “Craig, you know I live for adventure, but let’s leave and not come back until we are better prepared. The Library of Atlantis may tell us who the bad guys are and give us some fantastic new science, but I think you are out-voted.”

Andrea said, “I don’t even know if SHIP can be reprogrammed to act offensively. I suspect that given the choice of being offensive or not protecting the crew, the SHIP would go on the offensive. Let’s give me some more time and go to the colonies.”

“Okay, let’s run away to fight another day. SHIP, take us in the wrong direction for enough light years they could not track us. Put all your stealth capabilities on full power using every capability you can. Do that search for sensors that may have stuck to or in you somewhere. When you are sure there are no sensors and we are at least double the sensor range of that ship, then take us back to the colonies as fast as you can safely go.”

Three days by Earth time later, SHIP had found some sensors stuck to its hull. There were even a few cuts in the outer surface and one of them had a sensor inside. SHIP disposed of those by disassembling them to sub-atomic level.

Two weeks later, SHIP announced they were over twice any possible sensor range from Earth. "Which colony do you want to go to first?"

"Pangaea is the furthest away, but most of us consider it our favorite and Adriana has not seen it yet. We will go to Vulcan last. Hope you don't mind Jane."

"I am just Hua Ming a sub human woman on Vulcan. I would just as soon never return, but if we must at some time, I will."

"And I understand why you don't pick Pyramid," Adriana volunteered, "The people on Pyramid have it too easy. They have not had to carve out an existence on a wild planet. Almost everything was provided for them by the Martians. Pangaea and your crew have introduced more new science there than anywhere because it was needed on a wild raw planet. The scientists there are probably more open minded to developing a defense rather the soft life on Pyramid. I know nothing of Vulcan personally, but from what I have read and heard, I agree that they are the least desirable to help. Let's go to Pangaea."

# CHAPTER SEVENTEEN

# RUN AWAY TO FIGHT ANOTHER DAY

During the trip each person worked on what they could to prepare the colonies for a defense. Craig asked SHIP, "Could you build a miniature of yourself to spy on the mystery ship?"

"No."

"Could you build a ship with your stealth capabilities so we could get close without having to risk ourselves?"

"Yes, but it would not be fast and have to use more conventional propulsion. I do not know how I use gravity and anti-gravity, I just do it. The propulsion used by the colony ships is too bulky to build into something small therefore it would have to have ion propulsion. We would have to be within a few million miles and then communicate with the small ship using laser beam with delays. Maybe Adriana can program in some artificial intelligence. I know I am an intelligent ship, but I did not build myself."

"SHIP is right; I have a lot of experience in building in artificial intelligence. The small fighter ships I was forced to work on for Vulcan had artificial intelligence for long trips so the small crews could sleep enroute. They were even going to be capable of close in attack and evasion without the crew being involved. I did not complete the work as I did not agree with the whole idea, but I am sure that I can do it if SHIP can build the electronics as I specify."

"Easy. Just load your designs in and I can produce limited quantities in seconds."

"Can you build some sensors to attach to their ship to spy on them?"

“Yes. I could use their technology on them. Before I absorbed their sensors they planted on me I copied their design. The sensors would allow us to track them from one hundred million kilometers. That also tells me they were within Earth’s solar system watching for us. We were only in the solar system for a few weeks and nothing can transmit faster than light so just they would have to have been close by. I think we set off an alarm by sending something into the Library of Atlantis. That implies that they know where it is and I suspect they want to get into it also, but cannot. As quickly as they showed up, within hours of our peek into the Library, they must have been near Venus or the asteroid belt to get there that quickly.”

Jane said, “I need to improve my design for the beam weapon so it can be mass produced in the colonies in case they are attacked. I see no way to miniaturize it because even for cutting and welding Impervium it needs lots of electrical power and SHIP is telling us it cannot produce a small enough power source for faster than light travel so our spy ship cannot have a good weapon.”

“SHIP?”

“Yes, Dawn.”

“You analyzed the sensors they attached to you; did you also analyze the weapons they used?”

“Of course.”

“Can you reproduce them?”

“I detected what they were, but not how to make them. Regardless my programming will not allow me to kill intelligent creatures.”

“Adriana may be able to change that, but if not, you may be able to disable most of their ships capabilities.”

“Possibly.”

“So Dawn asked if you could reproduce their weapons so someone could use them against them. You said you did not know how to make them but knew what they were. What were they?”

“The torpedoes fired at us were anti-matter torpedoes. Each one had the equivalent of a thousand Hiroshima atomic bombs. They would not have disabled me, but could have done some minor damage so I just assimilated them as sub atomic particles which gave me the stored energy to do other repairs and the escape dash.”

“They used particle beams that were intense enough that I could not stop all of the particles and they did some of the damage. It would have done serious damage to their ship if fired at them instead of me.”

“The other damage was caused by projectiles that came at me like shotgun pellets made of anti-matter. I was busy absorbing the torpedoes and the particle beam and some of the tiny pellets got through. Ordinary matter could not have harmed my outer shell, but the anti-matter caused severe pitting. Some of the pellets were actually tracking sensors. I was able to absorb thousands, but not all of the pellets.”

Craig asked SHIP, “Which of these weapons would you recommend using on mystery ships?”

“Hua Ming’s or Jane’s, beam. Messing with large quantities of antimatter is too dangerous to the colonies and beyond Earth technology that you brought with you, but Jane’s beam weapon is totally safe to manufacture and relatively safe to use as long as it is controlled. Besides, since I did not know of its possibility, then they won’t either.”

“Not to rain on the parade, but Jane’s beam takes lots of power, which I guess means that it will have to be a ground based weapon, which means it will have to be able to fire through the atmosphere including heavy clouds, and thunderstorms." Craig said, "What will that do to its range?"

Jane defended it, "It is very powerful. I have never built a full power beam and have not tested it, but it should have no problem going out into space with enough power to destroy an Impervium ship, or at least put a good hole in one."

"I assume the beam goes at the speed of light."

"Yes, of course."

"So if you were aiming at something at the distance from the sun to Earth it would take eight minutes for the beam to travel ninety two million miles. Jane, that means in the time the ship you are firing at will have traveled nearly two hundred million miles while waiting for the beam to arrive. Even from Earth to Moon could take slightly over a second."

"Okay Craig, so the beam will only protect the Earth at close range, but then we want a defensive, not an offensive weapon. And, we need at least six per colony so there will be one in each direction of approach. That means having massive generating capacity around the planet. I would like to point out that the beam will spread some over distance to give a larger aiming area."

"As the beam spreads it may cover a larger area, but will be less powerful. Our colony ships and their nuclear power plants are not spread around each planet. They are pretty much concentrated in a relatively small area. So we would have to build other generators around the planet or in the case of Pangaea, we would probably have to put beam weapons on a ship in the ocean because one hundred eighty degrees from the colony there is no land. It would take multiple ships to do any good."

"I think all we should aim at is protecting the actual colonies. We simply cannot protect the whole planet."

“Craig,” Jane interjected. “I agree that we cannot set up my electron gun against Impervium ships, but we don’t know their ships are made of Impervium. While on Vulcan I almost finished the development of small Impervium fighter size ships armed with missiles and I was working on a small low power beam weapon. The intent was to breach the hull and fire a missile with a conventional explosive warhead into that small breach. I never released any of the designs or ever put them in the computer systems, but I think I have thought through the designs that will work. I just have to put them on the computer, build prototypes and test them. The small ships could patrol from orbit reserving their nuclear power if they spot something. I don’t think the Vulcan ships would have ever been able to travel at super light speed to get to another distant planet, but they should work quite well near the planet as patrol ships. I designed them for multiple take offs and landings. Unlike the old space shuttle, the Impervium makes it a perfect reentry vehicle with no heat tiles to replace or come loose. With small nuclear power plants onboard and a closed environmental system the only reason to land is to change the two person crew and resupply the water and food. Even most of the human waste would have been recycled for human consumption, but each conversion loses protein and vitamins and humans need. Consequently there would have to be a landing every couple of months. To get into space would have been through solid fuel boosters to achieve orbit or even escape speed and then ion propulsion for further flights. However, the problem with food and lack of super light speed prevents them from going too far from their home base.”

“So, the other colonies had nothing to fear from the Vulcans?”

“I wouldn’t go that far, Craig, they had plans for a larger ship to act as a spacecraft carrier. It would have been a copy of the colony ships, but smaller with a smaller crew. It would have super light speed to reach the other colonies and would have been carrying the smaller two man ships. I never even started the design, but others on Vulcan only had to copy the colony ship design and convert that design for a carrier."

Dawn asked, "How long would it take to build, test and produce a fleet of the small ships to protect Pangaea?"

"I don't know the capabilities on Pangaea. Two or three years minimum, I would think. I would guess four to six years to develop the industrial capacity to produce them."

"How many ships would be needed?" Craig asked.

"I have been thinking about that these last weeks and working through equations with SHIP. As a minimum, I decided at least thirty small ships. Only ten would be in space at any one time and that would be if we launched some Hubble style space telescopes to search for invaders at long distance."

Craig stated, "When traveling at super light speed, their ships will get there before the image on the telescope will. At ten times the speed of light, if you see them on the telescope millions of miles out, they would have been there weeks or months."

"There was a hole in my thinking, but they will have to slow down below the speed of light and then we can see them coming. SHIP has antigravity so, to use an American term, it can stop on a dime without us even knowing it stopped. But if their ships are like the colony ships they will have to slow down for a long time. They may be visible for at least weeks before they arrive."

Karen volunteered, "Therefore, the small ships don't have to be in orbit but can be on the ground ready to go like the old intercontinental ballistic missiles on Earth. However, we don't know they do not have antigravity. I think we need to get each colony started on building ships then go find the source and see what we can see."

Jill piped up, "I have been working with ship on drones that might be able to sneak up on that ship watching the Library of Atlantis. They obviously don't have teleportation or they would have ransacked the library thousands of years ago. If we could somehow find their home planet we might be able to learn more about them. The drones may be able to sneak up on their ship, attach itself to their hull and provide us telemetry data to tell us where they came from. As a minimum we may be able to eavesdrop on them. If SHIP can interpret their language we should learn a whole lot about them, like where they came from, why they fired on us, and what threat they actually pose to us and the colonies."

Dawn asked, "Why did they fire on us? We don't even know who they are? I thought mankind was the only warlike race."

Karen and Jane started to answer simultaneously, but Jane let Karen answer as the more senior family member, "It is the nature of animals to rule. Even sweet little deer have a buck that rules the herd. Herds of horses have a lead stallion. People are just sophisticated animals. Someone wants to rule the roost. Here, Craig. Sorry Craig, but we let you rule as the captain because SHIP made men the captain. The rest of us girls let you rule because SHIP chose you first and you chose us. Just like the head wolf leads the pack. We let you lead. Fortunately, you are a benevolent boss. You always listen to us as a good leader should. You let me make most of the decisions in my specialty, weapons and protection." You let each family member almost total freedom."

Sandy could not resist interrupting Karen, "But Vulcan's leader was not benevolent. He was a dictator that expected his desires to be law like the king's of Europe and the Middle East. He wanted to rule. You, Craig, don't want to rule. You were placed in an impossible position and recruited your crew to help you. The good of all mankind is your objective, not absolute rule. Jane?"

"Absolutely. On Vulcan, women are things to be used and abused. Only self preservation of his kingdom, Vulcan's ruler gave us some protection and a few freedoms, but that was to make his kingdom survive. The women became more important for breeding more serfs. I was one of the few that were actually allowed to excel at something other than breeding and satisfying the men. My time in that capacity was shrinking. Within the year, I expected to become just another member of the harem. Yes, there is always a leader that emerges in any set of animals. Hopefully, our colonies of Pangaea and Pyramid will be run like democracies and not become dictator or kingships. Vulcan will never change. There are already men fighting to position themselves to become the next master of Vulcan."

Karen added, "Most of the history of mankind has been about some strongman trying to take over other peoples land, money, etcetera. Apparently, it goes beyond humans."

Dawn argued, "Apparently not all intelligent creatures. It would appear that Martians left Pyramid behind for us to take over if we ever made it to space. That was no mean accomplishment, and someone, if not the Martians, left SHIP in orbit around Earth to bring us to this point. That is doing good."

Craig agreed, "You're right, Dawn. We are now thinking that someone more advanced than the Martians left mankind the ship. It appears the Martians left Pyramid for evacuees of Earth, whether it was for the people of Atlantis that were left behind or humans, we may never know. The United States did many good things for the other people of the world without expectation of return, like rebuilding Japan and Germany after World War II."

Karen chimed in, "Not always. The United States took the land from the Indians that were there first. However, I agree, the United States matured and later tried to help the Indians and provided a lot of freebees like free college. I would say as democracies learn, they get more beneficent. However, Plato said that Atlantis was conquering Egypt way back in history with ten thousand of their men taking on hundreds of thousands of Africans and winning, but they retreated to Atlantis and then disappeared from Earth. According to Plato, they were recalled to Atlantis because of something bad happening at home. If the story is true, Atlantis was not a friendly people either. Maybe Pyramid was prepared by the same intelligent creatures as built SHIP. Whoever built SHIP is far beyond whoever they are that attacked us while trying to get into the Library of Atlantis. Maybe, like the United States, they went through a conquering period and then mellowed just like the United States."

# CHAPTER EIGHTEEN

# FORTIFYING PANGAEA

SHIP could not go full speed in the galaxy due to the number of suns, planets, asteroids, and free rocks, but at even that incredible speed it took three months to arrive at Pangaea.

The family worked hard during those months. Karen for tactics, Dawn for astronavigation, Andrea for computer programming, Jane for weapons, and Jill for engineering locked themselves in the dining room for hours of strategy planning. Craig stayed out of the discussions except to step in occasionally to keep up with what they were thinking and add suggestions or criticisms. Sandy took those times to take Craig to the bedroom alone. When the science discussions got beyond Karen's education level, Karen and Craig went to the master bedroom, while Sandy went back to studying the zoological specimens they had collected. They all took breaks for meals together in the dining room. For hours each day each girl worked on their own specialty to develop defensive measures against this new unknown race that had fired on them near the Library of Atlantis.

Karen came and went from their meetings and kept Craig posted on what the others were working on. "The plan is to develop a small electron beam weapon to disrupt a very small point in an Impervium ship and then fire in a miniature nuclear missile that would explode inside the shell. By using a very tight beam, Jane thinks she can make one powered by a small reactor in a fighter size spaceship like she was developing for Vulcan. On Vulcan the plan was to cut up the Impervium from a colony ship to build the small two person spaceships and since Impervium works as a radiation proof reactor they were going to cut up the plutonium from one colony ship to make many small reactors for space propulsion. She never turned over the designs to Vulcan, but she has them in her mind and has transferred those designs to SHIP. Computer simulations appear that both the reactors and the beam weapons will work. The beam starts out as nearly microscopic but will spread to two inches at ten thousand miles of empty space."

"The missiles are being developed by Jill to be under two inches in diameter. She has simulated a new very powerful, long lasting solid fuel for the missiles. The missiles will be twenty feet long to hold enough fuel to accelerate the missiles to one hundred thousand miles per hour. The missiles are very small and coming head on, hopefully, will not be detected by enemy ships. The beam will not be used until just before the missiles strike and weaken the spaceship shell to allow penetration of the missile."

"Adriana has designed a miniature computer guidance system that once the enemy ship is sighted and the missile locked on the missile should fly true until impact. Because when in space the speed of the missile will not diminish for years unless it hits something. The missiles will have small rockets for guidance if the enemy tries to evade. If it loses the enemy ship it will self destruct in a non-nuclear blast to just pulverize the plutonium to become background radiation of space. Dawn has given Adriana extensive help in astronavigation to help with the guidance,"

"Sandy is anxious to find out more about the enemy. Are they humanoid or fish or what? Is there any way to reason with them? She is pushing to get some small missiles equipped with tracking equipment to find their home planet or their base at least."

"I am against that because that could give them warning to look for small missiles coming at them. Once they know about them, they may look for them and start finding them before they can do their job. In addition, tracking equipment will only work at the speed of light so at a paltry million miles there will be a delay in getting the signal. At several million miles they could have changed course to make it impossible to follow. So what do you think, Craig?"

“I like the idea of tracking them down. Maybe we could use something that would decay for months that we could follow the traces of. If it comes to war, we must be able to find their source; otherwise they could just launch attack after attack at us. I would like for the girls to consider Sandy’s suggestion and the idea of them leaving a trail. I agree with you that we must first use the missiles offensively if they attack us. We have nothing against them unless they track us down and attack us. Then we will have to fight them and try to destroy the attacking force. Maybe if we can make them retreat we could then use the trackers to see where they go.”

“Why don’t you join the rest of us in the dining room and tell them?” Craig and Karen joined the rest of the family and they decided to work on the trackers after perfecting the designs for the weapons. They would not be able to build the hardware and test it until they reached planet fall on Pangaea.

Craig asked, “Is everyone in agreement that we will do the development and testing on Pangaea first and then after testing is complete go to Pyramid and have them duplicate it?”

Adriana objected, “I think we should start the development on Pangaea, but then we need to take the sum of our knowledge and then, at least some of us, should go on to Pyramid to start a parallel development there. It may take several trips back and forth to take lessons learned between the planets. It is possible that the enemy might have some knowledge of Pyramid already, so they could be attacked first. After all, the Martians did not know of Pangaea, but they built up Pyramid. In fact, looking back, maybe the Martians built Pyramid for themselves and then evacuated it without occupying it because it has already been discovered by the enemy. Don’t get me wrong, from what you have told me about Pangaea, they seem to be more pioneering than the laid back Pyramid people. Most of the work of building a civilization was already in place when the colony came to

Pyramid so the engineers may have been more complacent, but the selection of colonists was the same for both planets."

Karen asked, "What about Vulcan?"

"Remember I am from Vulcan. I deliberately held back providing war weapons to Vulcan while I was there. I would not want to contribute anything that could be used against the other two colonies. Maybe after Pyramid and Pangaea are secure, but that will be years. If they get attacked in the meantime, so be it. I would feel sorry for the women there, but I don't know of any men that deserve to survive."

"Okay, Jane, sorry I asked." Karen said.

Craig continued, "Back to Adriana's suggestion. I agree. I think that Jill, yourself, and Jane need to stay on Pangaea, but myself, Karen, and Sandy should go on to Pyramid immediately to get them started on the development also. Dawn, what about you? Should you stay behind or go on to Pyramid?"

"I should go to Pyramid. You, Sandy, and Karen are not up on your astronavigation. I am sure that SHIP knows the way, but I think I have a better understanding of the new weapons and that Jill, Adriana, and Jane have all the knowledge needed for Pangaea. We all need to stay on Pangaea for a few weeks to get the effort organized before we leave. It will take three months to get to Pyramid, at least a month to get the effort started on Pyramid and then three months back. Pangaea will have had six to ten months start on development and testing by then so there may already be changes that need to go back to Pyramid. Again, by the time we get back to Pyramid, input the changes and get back to Pangaea then another six to ten months, and so on. So there will be a lot of travel. We don't know if we have weeks, months, years or if we never get found and attacked. At some point we need to go back and explore the Library of Atlantis."

"As always, I agree with Dawn. Jill, you know about being isolated from SHIP and you are familiar with Pangaea. I think we will need to stay at least a month on Pangaea to get things started and acclimatize Jane and Adriana, but a word of concern for the two of you. SHIP keeps your age at your prime age. It keeps you in top physical condition without exercise. It can heal most wounds you might get outside the ship. Broken bones are no sweat as long as you don't bleed to death or get a fatal infection on the planet before returning to SHIP. When you teleport to SHIP, SHIP will remove any harmful infections, diseases of any kind, and put bones back the way they should have been based on your best DNA. However, you could die while outside SHIP. I don't know that SHIP could replace a missing limb. Any other person on Pangaea can be easily replaced. We now know that you are replaceable too, but a lot went into selecting you for SHIP so finding a replacement for you would be difficult and might be impossible at this time. Therefore, you should take no chances while on planet. Stay in safe zones; make sure all experiments are safe. If you must have someone in a dangerous situation, let it be someone from Pangaea, there will be no shortage of volunteers."

"Actually, before we break up this meeting, I have some serious questions about our weapons. Jane, what happens when the beam hits a more conventional metal like aluminum, magnesium, or steel alloy?"

"I did some testing on those with my low power beam on Vulcan, it vaporizes instantly. "

"What about a ceramic spaceship?"

"Good question. It has not been tested, but ceramics are not metal and typically have a ridiculous melting point. It may not work, but why would you think of ceramics?"

"Because the space shuttle had ceramic tiles as a heat shield, so why not build a whole ship from ceramics?"

"We'll have to test it. Regardless, the miniature nuclear warheads in the missiles should take care of anything but

Impervium and the missiles exploding inside an Impervium hull should get rid of any life therein."

"Craig, I would like to stay on Pangaea while you, Karen, and Dawn go to Pyramid. You don't need me there, and I could stay and help in some way on Pangaea and do some more study of the animals there. I still don't know why cats seem to be so dominant there. I am worthless for physical science, astronavigation, and cannot help on Pyramid, but I can help on Pangaea."

"Okay.

A week later they teleported to Pangaea. They were given quarters for their stay and assembled engineers for the development effort. They were surprised to learn that Pangaea had been able to start producing their own Impervium in small quantities to manufacture tools for housing construction, farming, and repairs. They had completed the oil pipelines so that petroleum was able to produce electricity, heating and cooling for their buildings and fuel for equipment. They were also able to produce additional batteries for additional equipment. Their primary concern was expanding their farming. Their population was exploding with an average of eighteen children per family, but the children were not able to contribute to food production. Each woman had produced an average of three children in the intervening years since planet fall with some having four and five children. That was the one of the main reasons that the colonies had six women for each man, produce children while the women did not need as much food as the men did.

# CHAPTER NINETEEN

# REUNION ON PYRAMID

There was no way to radio ahead, but they went into orbit before landing to spend a day circling the planet and observing changes while they were gone. During this time they did talk with the ground to give them a chance to arrange quarters and assembly the engineers and mechanics necessary to build prototypes of their designs. When they landed Estera was leading the group to greet them and introduce them to the assembled team.

"Estera! Good to see you." Craig exclaimed. Karen, Dawn, Sandy, and Jill crowded around and hugged her. Jane and Adriana hung back as they did not know Estera.

"So, Estera. How have you been here on Pangaea?" Dawn asked.

"Good. Very good. I have found a family that is compatible with each other. I have two children of my own, but one woman has had four. I have been very healthy. My eyesight and other infirmities that were fixed by SHIP have not come back. I am still a ninety pound weakling compared to the rest of you, but two children have made me weigh one thirty now. I need to get more exercise to lose the baby weight."

"So what have you been doing besides having children?" Jill asked. Jill and probably the others were noticing how Jill looked like she had aged ten years in the last four on Pangaea.

"Oh, I have been busy. Bet you did not notice that the city has not grown as much as you would think from the population growth. We have actually built a lot of new structures, but they are covered by vegetation.  I have been busy as a biologist. Because everything was being built from scratch I took a lot of information from the colony ships computers about the research in natural cities and started. This is something we studied in college, but there was mainly talk and no action on building natural cities. The vegetation covering the buildings helps in many ways. The roof collects rainwater and keeps some of it from running off into the streets reducing the need for massive street drainage. The water is filtered naturally through other plants to be drinkable and water the other plants. This reduces the number of wells we have to drill for water and some of the chemical treatment of that water to remove chemical impurities in well water. The vegetation cools in hot weather and insulates in cold weather. Some of it is even edible and quite tasty. Each house now has plenty of spices for not only themselves, but to share with the community. In case of storms the buildings are reinforced with the plant strength added to the building strength."

Dawn replied, "Wow! You have been busy. Good thing you decided to stay here. How is life outside your work?"

"It has been great. I was never an explorer and felt useless on the ship. The adventure here is all I need for excitement. I do miss you all and of course, Craig, but I have a good husband here and I do like my family women. We take care of each others children. That has allowed me the time to work with my plants."

"It works great as camouflage. We did not notice the expansion of the cities from space. That will be a great help." Karen stated.

"Why is camouflage from space good?"

"Because we were fired on by someone from another world when we were in Earth orbit trying to explore the Library of Atlantis. On the way back here we spent the whole voyage working on weapons to counter a potential attack. The better we can camouflage the cities the better we can hide from a space attack."

"Oh! I am glad I stayed here. I could not have helped prepare for an enemy and I guess my work here had a duel purpose to prepare for an attack as well as just building green cities. Is there any eminent danger?"

"Not that we know of. SHIP had let down its stealth defenses since we did not know we needed to hide from anything and the defenses required power that SHIP thought it was wasting. After getting away, SHIP went into full stealth mode and we returned here by a round about path to make sure we weren't followed. We came here first because your people have had to build everything from scratch and we figured that would make your engineers better at inventing something new. However, Pyramid may already be known to the enemy and Craig, Dawn, and I are going there as soon as we get the defensive development going here. Then we will go back and forth transferring technology from planet to planet."

"What about Vulcan?"

"We think that it is not as hospitable as here and less likely to be found by the bad guys. Plus, Jane, our new Chinese crew member, is from Vulcan. They wanted her to build weapons to conquer the other colonies, so until we have Pangaea and Pyramid defended, we don't want to help Vulcan that might use this knowledge against us. Once we have defenses, we might go help them. In the meantime, if the enemy finds Vulcan first, maybe they will think they got us and will quit looking."

"I guess our decision to isolate them on their own planet was a good decision."

"Most assuredly."

Sandy took Jane and Adriana on a tour of Pangaea explaining why the colony ships were located where they were and how they were interconnected. Jane was amazed at the freedom of the women doing all sorts of jobs and free to roam the planet when not working. Jane stated, "I see no militarization anywhere. Other than hunting weapons I see none."

“There aren’t any other weapons. We still hunt the buffalo type creatures for food and have to defend ourselves from the large cat type creatures, but we have no need to defend against people, until now with the threat from an unknown enemy.” Sandy replied.

“The women are everywhere and really outnumber the men. I understand this was intended for all the colonies. Why did your family allow the Chinese to take weapons into space and only enough women for their amusement?”

“The alternative was to leave the Chinese to perish on Earth. We tried delaying the delivery of the spaceship engines and other technology. We threatened to just leave them on Earth, but Craig could not stand to let them just die on Earth. We had found Vulcan, and while not friendly, was habitable. Pangaea was like an unspoiled Eden. Pyramid was already partially developed and just waiting for people, so we isolated the Chinese on Vulcan for the protection of the other colonists.”

“I have another question. Are women prohibited from reading certain material or have access to the computers on the colony ships?”

“Absolutely not. Women not on Vulcan have access to the total knowledge of the computers which includes everything digital available when we left Earth from scientific articles to novels, movies, and old TV shows. Not only that but the women are probably more involved than the average man. Men can do some of the manual work too difficult for the women, but women are more adept at verbal skills. However, when it comes to spatial concepts held by the computer, the men seem to excel. Neither sex is prohibited from doing anything that does not hurt others.”

Adriana said, “It was the same on Pyramid. Women could do everything but if you could not produce your share of the children, you did not fit in nicely. You probably know that my accident messed me up for having children. It’s not like I was outcast or anything, I just didn't feel like I was fulfilling my job as a family member when I could not have children. I sort of isolated myself along with a few other women that had something go wrong with their reproductive system. According to SHIP, I was healed by the SHIP, but now I am not concerned with having children or not. There is a mission to do. I think that feeling is shared by Dawn, Jill, and Karen. What about you, Sandy?"

"I think Dawn, Jill, and Karen are not only committed to the mission for as long as it takes, but they love Craig and want to stay with him through thick and thin. I think Dawn and Craig are like an old familiar married couple, in fact, they are actually legally married by Earth law. Jill is scientist first, but has physical needs that Craig can satisfy. She is not only the best scientist, but probably the most committed. Karen seems aloof. She is fiercely independent. I think Craig is the only man she has ever been attracted to and does not want to change. I can't see her wanting to settle down with children and be a mother first and give up her independence and exploring nature. With her, it is not love of Craig, or sexual attraction, but the adventure and Craig is there to share everything with her including the semi-regular sex. I know I am different than the others. I think Craig has more feelings for me than the others, but I don't mind sharing him. I was the last of the original crew members because Craig was concerned that our physical attraction to each other was too overpowering. Once I met Craig, I did not want anyone else. Craig says he stayed away from me because he was afraid that the physical attraction would interfere with his decision making and he needed to find five other women first. Dawn was the first by far. Jill was Dawn's friend and attracted to Craig from the start, lust not love. Shelby was a chemist. Craig needed someone more capable than him with weapons and living off the land as protection on planets for the family. Then Craig came back for me after a two year wait. I think Craig loves me as much as I love him, but just like I have to share him I think he has to love the other women too, I would like children some day, but Craig and I are not getting any older. Therefore, my motherhood clock is stopped. I am enjoying just being with Craig and being a part of this adventure. Maybe someday, his mission will be complete and we can settle down, raise children of our own and get old together. Right now I have enjoyed the last ten years together."

"But you and Craig are no longer the teenagers you were when you first met. I am fifteen years older too and I was twenty one when I left Earth so I am now an actual thirty-six years old. I have only been on SHIP for a year and I have teleported so few times I still cannot teleport fully clothed like you can. How long will that take before I can teleport freely?"

"It took a year for me. I enjoyed teleporting being nude with Craig. Of course, the original family members teleported a lot during that first year. It only took Dawn six months, but they were teleporting at least once a week. I suppose that means that SHIP must teleport you nude for between twenty and thirty times before it can keep you separate from your paraphernalia. It would not do to mix up your DNA with a fly like in the movie, "The Fly"."

"No, I guess not. I really like Craig and believe in what we are doing, but I am more like Estera. I am enchanted at the wild nature of the planet versus the built up areas of Pyramid. This is like Eden in the Bible. It is so wild except near the colony ships and the roads between them."

"You haven't explored the other continents on Pyramid yet."

"We never explored much beyond our own continent. There was so much of that continent we never even explored let alone go looking for other continents."

"They are vastly different than the one you are living on. We did a cursory investigation of all the continents. Pangaea only has this one continent and no one has explored very much of it.

Think in terms of living in France and not knowing the rest of Earth is there. Imagine no one living in Germany, or Italy, or the Americas. There are no humans living anywhere on any of the planets except near the colony ships. We only looked for a benign area with plentiful resources for a colony and to see if there could be other humans livings on the planets before we selected the landing zones for the colony ships. Even Vulcan has not been explored. In fact, we did not even explore much of it except to determine the best quickly identifiable landing zones. Take Pangaea. It appears that the north and south are very similar with cold at the poles and hot at the equator. The temperate zones of north and south were similar. Since it is a new planet, what makes north or south? We arbitrarily chose the location of the colony."

Jane asked, "Why did Estera decide to stay here on her own, join a family and have children. She seems very committed to science. Why didn't she stay with SHIP? She already looks older than I would expect."

"Estera joined the crew as a political thing. Her father was the head of the United States House of Representatives Ways and Means Committee. That means he had a big say in whether the United States government provided funding and facilities to support Craig's mission. He believed in Craig's prediction of Earth being destroyed. In exchange for his political support Craig agreed to take Estera off the planet. Estera went along because she too believed in the prediction and had helped convince her father to support the mission. Craig was obligated to take Estera. Estera was plenty smart enough but did not meet Craig and SHIP's physical standards for a colonist so she would not have been chosen except for this deal. Estera was never in love with Craig. Estera could not do much more for the mission once the colonies were established, but saw Pangaea as an opportunity to do more by staying here. It also gave her the opportunity to choose her own man versus her father brokering a deal."

"Shelby and Craig were never in love. Craig needed Shelby's expertise and Shelby needed a way off Earth. Shelby is also a big science fiction fan that had always dreamed of being able to leave Earth. Shelby and Craig did have occasional sex, but it was more obligation than love making. Shelby chose a more developed planet, Pyramid."

"Karen and I are the only ones that did not go to the University of Oklahoma. Karen was from Colorado and I was from California. Yeah, a southern California beach girl."

"Why have you decided to stay behind here on Pangaea? It sounds like you want to stay with Craig."

"It will only be for a few months. I am the zoologist of the team and I have not done everything I can for the breeding of the animals we brought from Earth. All the test tube animals survived and are now breeding on their own, but I need to study them for ill effects of living and eating on a strange planet. I also need to study the native animals more. I am only vaguely familiar with all the variety of cats, from huge lion size to small monkey cats. There are insects here that have not even been identified. Dedicated members of the family on SHIP have potentially forever. A few months apart here should not be a problem."

"Won't you be jealous of Dawn and Karen being alone with Craig for maybe half a year?"

"That is hard to be jealous when we have all been living together for fourteen years now. Dawn has been with him for nearly twenty years, but they are only in their twenties. SHIP seems to have let us age to a prime age for each of our DNA sets. Adriana was new and it reduced her physical age fourteen years. It is only for six months. If any of us were in the military on Earth a six month or a year's separation would not be unusual. No, I do not want to be permanently separated from Craig, but if we were to ever want to settle somewhere it would be here on Pangaea."

After a week of traveling around Pangaea, Sandy took Jane and Adriana back to the central city where the engineers and mechanics were gathering to attempt to build small spaceships and the new weapons. Jane and Adriana had an understanding of the uniqueness of Pangaea and knew that this was the first place to defend.

There were four teams: small spaceships, miniature long range missiles, miniature nuclear warheads, and the long range beam weapons. A fifth team was building the original beam device to cut and weld Impervium to reuse the Impervium from one of the colony ships.

It took six months of intense work to build their first small spaceship and test it on a short spaceflight. They used chemical rockets to get it into space and then tested the nuclear ion drive for a short acceleration test. The acceleration was only a partial gee versus the colony ships had maintained over a one gee acceleration for months.

The beam weapon had been tested in the atmosphere from an aircraft at over three hundred miles, but without it going into space they only had computer simulations of its range in space. At three hundred miles it could instantly vaporize a small hole in a piece of Impervium. However, the dispersal of the beam by the atmosphere reduced the power. In space it would be even more effective. Any beam weapon was limited to the speed of light. Detecting an approaching space ship was limited to a few million miles with luck so the beam weapon had to fire through space for a million miles and still have the power to vaporize a small hole.

The small nuke warhead was perfected on computer and a few built to that specification, but no one wanted to test it on the planet to spoil an almost radiation free atmosphere. The family decided it would wait for the test of the new miniature missile where it could be sent hundreds of thousands of miles out from Pangaea for testing.

Jane had what she thought was a perfect guidance system developed for the missile. If they could detect an incoming spaceship, they could target a missile and it would unerringly find its way to the spaceship at near light speed.

The missile was easy to build, but the fuel for a near light speed flight in space was another story. Jill had been working on a new chemical solid fuel that far surpassed anything that had been available on Earth. It would accelerate the missile at twenty or more gee and keep burning for days allowing the missile to achieve near light speed in only fifteen days if the computer simulations were correct. The two inch missile would be packed with the solid fuel. There were small additional solid fuel rockets for terminal guidance once the missile approached the enemy spaceship.

Finally after eight months a small spaceship lifted a missile into space and fired one off with a miniature nuclear warhead. At a million miles out, the warhead was detonated. The guidance system had unerringly guided the missile toward a distant star. The rocket motor had worked better than planned and the warhead had exploded with an estimated one hundred tons of TNT from its two pounds of high explosive that set off the small nuclear explosion.

The team that started the small beam weapon that seemed to work started working on a super ground based beam weapon. Pangaea had plenty of electrical power for such a beam weapon. It would be a last ditch weapon if an enemy ship got past the small patrol spaceships.

They were still working on a way to power the beam weapon from a small space ship when Dawn, Craig, and Karen set off for Pyramid. They took all the most current designs with them as well as working models of one of the small spaceships, a couple of missiles and warheads, and a small beam weapon. They also had a list of the top scientists on Pyramid that Adriana had made up for them.

Without scientists on board, Dawn, Karen, and Craig entertained themselves on the trip by reviewing what they would do on Pyramid, what they had done on Pangaea, watched old Earth movies and had sex.

By the time they arrived at Pangaea three months later. SHIP was running as fast as was safely possible between solar systems in the outer belts of the Milky Way Galaxy. It would be impossible to run that fast nearer the crowded center, but if outside the galaxy even SHIP did not know how fast it could go.

They wasted no time surveying Pyramid but immediately teleported down to the pyramid in the middle of Pyramid City. It did not take long before the planet government was there to meet them. Craig insisted they go into a large meeting room so he could address the available government leaders.

Shelby was one of the leaders that showed up. As a past member of SHIP and a key member of getting their oil refineries and chemical industry started, she was asked to be the lead scientist for the planet. She ran up to Craig and gave him a hug and long kiss. "I've missed you. I am very happy here, but I have missed you terribly, but more on that later."

Craig told the assembled crowd what they had experienced with the attack on them and what they had done to date. He went on to say that they were planning to stay on Pyramid long enough to get them on a par with the Pangaea defenses. "Pyramid will take less time than Pangaea took because we brought working prototypes and all of the design data to date. Adriana that went with us last time we were here stayed in Pangaea to continue the effort, but has given me a list of the scientists and engineers that she thinks should be on the team. This is of prime importance because it may be essential for the survival of our colonies. We certainly cannot impress an enemy with numbers because we only have a few thousand people. We have to assume they have been space faring long before us so all we can hope is that we can surprise them with something new of our own. Once Pyramid can match the capability from Pangaea we will take your lessons learned back and get an update from them and then return with those updates." We have decided we will sleep on our ship, but spend long days with you here on planet. Tomorrow morning, hopefully you can have your specialists assembled into 5 teams for each of the technologies we have so far: A beam weapon that can instantly vaporize a small hole in Impervium, a small missile with good range and speed, a miniature nuclear warhead, a guidance system for the missiles, and then small fighter size spaceships to extend our defenses away from the planet. Once a team finishes we will then work on closer-in defenses of a more powerful beam weapon based on the planet and powered by your power grid."

They went out to the open area near the pyramid in the center of Pyramid city. Shelby asked, "I left the SHIP's family and I really don't want to return full time, but do you think SHIP could transport me there to be with you, Karen, and Dawn tonight?"

"SHIP, how about it?"

"Yes, I can teleport her. Since the others were left on Pangaea you will still be under the maximum and I am thoroughly familiar with her subatomic makeup."

SHIP transported the four of them back. They ate around the dining table. Shelby could not stop talking, "This is so great to be with all of you again. I have missed you all so much. Don't get me wrong. I hope you can understand, but I like being on a planet and having the open landscape around me. I have a family. There were a number of families missing a woman. Like, Adriana, once she had her accident and could not have children, she moved out and joined other women that for whatever reason did not like the expanded family with six women and one man or could not have children. That left families missing one or more women. I have found a family that I am happy with. I am compatible with the other women in the family, but it has only been a little over a year and I was with you for a number of years."

"Sorry, Craig, I was never in love with you and thought that I would like being a crew member on a spaceship, but I did enjoy it here. It just got claustrophobic after the years we were here. Whereas Karen wants the outdoors, I was always afraid in an untamed land. Here on Pyramid is like living in a small Earth city. I am comfortable here on Pyramid. The food here on SHIP is superb compared on Pyramid. SHIP knows what we want, what spices we like, everything is like a dream gourmet meal. SHIP chooses what we really want subconsciously versus what we might think we want to eat. There is so much more variety."

Karen interrupted, "There is something you aren't telling us. Out with it."

"I don't feel I have that right."

"SO. Out with it already. You were a full member of the family for years, nothing will surprise us."

"Okay, let me explain. The colonists have all aged nearly fifteen years now since we left Earth. My Pyramid husband was twenty-five by the time we landed here. SHIP kept me physically at twenty-three. That would have been fine. But now my husband of less than one year is forty and I while technically I am also near that age, physically I have only aged a year since leaving the ship making me twenty-four. Mentally, I am far ahead of anyone in the family so that is not the issue, but physically I am between generations. Bill, my husband here can barely keep his five long term wives sexually happy and I miss having sex with Craig who is still physically in his twenties and I presume still able to perform at that level. I want to sleep with Craig while he is here."

Dawn and Karen started to answer simultaneously. Dawn said, “Karen, go ahead.”

“Shelby. Of course. Dawn and I have had plenty of sex with Craig this last three months and we are used to sharing him with a total of six women. You can’t have him exclusively for the months we are here, but I think we have shared him with you for years before this. I think a week now and a couple of times a week would be perfectly within bounds. Besides, we will be busy with the new weapons here on Pyramid and might be too tired ourselves. However, keep in mind from time to time we will have to have our share.”

Craig said, “You surprise me, Shelby. I always felt that you felt obligated to be with me as a mate during our adventures and maybe for your own satisfaction. I am flattered that you want to sleep with me. What about Bill, your husband?”

“Things are not like they were on Earth. Men have no need to go find wild women when they have their own small harem. It is not that unusual for some of the women to spend a night at another house. Men occasionally switch harems for a week. It is not unknown that men make a permanent trade of families. It is kind of like the free love we read about that happened with hippies in the nineteen sixties.”

“I told my Pyramid family that I would be staying with the SHIP’s crew during your stay here even though that might be months.”

Dawn asked, “No pregnancy yet?”

“No. There is nothing physically wrong with me and my housemates have an average of three children with one having had five children. I think Bill is just not that interested.”

“But, Shelby, you are a beautiful young woman. You have a fantastic figure. I think you are very sexy.”

"But Craig, you know that you spent a lot of your time with Dawn, Jill, and Sandy. There was no question that Sandy and you have some great physical attraction. Not regularly, but I know you and Karen got it on pretty good every week or two."

"I'm sorry."

"Don't be. I never gave you the attention you deserved. You were actually good in bed. I have never lusted after Bill. He is just too old. He is a really nice guy. In physical age though he is almost a father figure. That may be why I have not gotten pregnant. What would SHIP do if you got me pregnant?"

"SHIP?"

SHIP replied over the intercom so they could all hear the answer, "This is different. When on the mission it was essential that there be no children here. That would still be the rule. I cannot teleport a pregnant woman because there could be a mistake mingling subatomic makeup of the child with the mother. Therefore, if I allowed Shelby to get pregnant here, I could not teleport her down to the planet with even the very beginning of a child. Therefore, I have never allowed any of Craig's women to get pregnant. The reason that I could not allow a pregnancy to start here is because it would be an impossible conundrum. The crew is limited to six women and the captain. I cannot teleport a young child that has not been physically mature … at least late teenage. I certainly could not teleport a pregnant woman. Therefore, if you and Craig wanted to have a child it would have to be on the planet."

"Actually several very early pregnancies were terminated by me before teleporting Craig's women here. It did no harm like an abortion. I just did not teleport those few cells. Fortunately, none got pregnant until after I had a good record of your subatomic makeup so I could teleport the woman without teleporting that portion that could have been a baby."

Karen said, “That’s gross. Are you telling me that I might have been pregnant with Craig from one of our trysts on a planet?”

“Yes, but by not teleporting the beginnings of a child I protected the crew. I could not allow more than six women and one captain. Not eliminating the child would have meant leaving you on the planet to perish and losing an essential member of the crew that Craig thought necessary for the mission.”

Dawn said, “Did you ever think about how I would feel to have you take a pregnancy away?”

“No. It was best for you and for the mission. You are still physically in your prime and as long as you teleport regularly you will remain in your prime. If you ever declare the mission over or voluntary leave the crew for any reason, you will still be physically at your prime. If I had allowed teenage pregnancies there might have been problems due to your young age. What about having to leave one of you in one of the dangerous areas prior to the colonization of that continent? You would not have had the baby because you would have died. Even Karen could not have survived alone with no human support?”

Craig said, “Okay, SHIP. What if I decided to stay on a planet with one or more of my women to have children and then something came up that I needed to return to you for another mission?”

“I don’t know. My programming limits me to one male captain and up to six wives. There is no allowance for children. You know which women I have memorized the subatomic structure of. I could teleport any of you back here at your current physical age healing most infirmities you might have collected on planet such as broken bones poked out eyes, loss of hearing, disease that has not killed you yet, cancers. But Adriana and Jane could not be teleported without being nude and with you. Karen, Dawn, Shelby, Estera, Jill, and Sandy would be returned to their prime age.”

Karen had been upset at the idea of having lost several pregnancies without knowing it. Then she was upset at the very idea that she had gotten pregnant while on one of the planets they visited. Then she laughed, "So, what you are saying if our little family were to settle on Pangaea and have children and get into our sixties, we cold teleport back here and be younger than our children."

This changed the whole thought pattern of everyone. That would be comical almost. They could literally have children who could have children and by repeating the trips to SHIP could mean that they could be younger than great grand children who had died of old age on a planet. It made everything okay. Dawn let SHIP off the hook, "SHIP, I was originally horrified to think that you had been killing children shortly after conception, but Karen has put it in perspective. I don't think any of us would have wanted to abort the mission to have a child at the time. It is not like we are getting too old to have children. Theoretically it might be possible for one of us to end up having a child by an ancient ancestor that had aged normally while we had come and gone from here."

Shelby said, "Okay. Let me put out this idea. Let's have sex on SHIP and keep me non-pregnant until this portion of the mission is over, but Craig, before you leave to go back to Pangaea will you try to get me pregnant before you leave?"

"That is a big question, isn't it? It is one thing to satisfy those physical needs and try to keep six women happy, but to leave children behind without a father."

"My husband, Bill, would be a good father. It is just that I don't know if I will ever get pregnant with him when I have to share his slowing urges with five other women. He really is an intelligent, strong, nice man who would like to think he fathered a child with me. He need never know it was you. In fact, without a DNA test how would we know unless SHIP tells us. I might get pregnant with you and not know it until you have been gone for awhile. When you leave, I will make the effort to seduce Bill daily for a few weeks. Then without a DNA test none of us will know, except I will have a child to fit in better. I do not want to be like Adriana and feel isolated because of not having children."

# CHAPTER TWENTY

# FORTIFYING PYRAMID

It took four months for Pyramid to build the industry needed to catch up with Pangaea in planet defenses. Pyramid had not had to manufacture much before since it was so built up already. They still had empty housing so manufacturing consisted of gradually adding wiring to additional houses, making windows and doors for the stone houses. Their engineers and mechanics had gotten lazy and with eleven years out of various schools on Earth they had forgotten much of what they had learned on Earth.

Finally, Craig, Karen, and Dawn were planning on leaving when Shelby stopped them. "I think I have invented a much better solid rocket fuel. It should increase the acceleration to forty gees and the time to burn to forty days and that is just the missiles. The space fighters can't accelerate that fast because of the human pilots, but as a result they could maintain their full acceleration for months with maneuverability limited only by the human pilots."

Dawn asked, "How long do you think it would take to build a missile to test this new fuel?"

"For manufacture and test, no more than a month."

"Craig, lets wait for the test and then take this new fuel back to Pangaea. If the fuel is really that great that means we could work on elimination of the low powered ion drive saving our limited plutonium for warheads and peaceful use in the colonies."

Shelby then laid the big one on, "While staying here another month, what would you say to the idea of Craig and I living together for a month on planet to see if we can make a baby?"

Dawn said, “I am in favor of it. Craig would have a child to survive him.”

Karen laughed, “Or a child to survive. The child will grow up and get old while Craig is still flitting around the universe.”

“Shelby, you have been a lot more enthusiastic lover these months while we have been here. Are you sure this is what you want?”

“Yes. Adriana told you what is like to be a childless woman here. They don’t ostracize you for not having children, but you just feel you don’t fit. It would make my husband, Bill, proud to have fathered another child and to have fathered a child with his young wife. It will make a difference to me to not be childless. It will also be a private piece of pride to know that it is Craig’s child. I did not give him enough credit when I was a member of the SHIP’s crew. I did not really let myself enjoy the sex then. Now that I have been here for nearly two years and these past months have been the only time in my life that I enjoyed sex.”

It only took a few weeks to build a test missile that worked great, but using the solid chemical fuel for the space fighters was another story. The fighters needed to be able to vary their speed. This was taken care of by having multiple rocket engines to selectively ignite. However, because of the energy density of the new solid fuel it was seldom necessary to use all the fuel in an engine. Therefore they had to develop a way to shut off a rocket engine only partially used then re-ignite the fuel when needed. SHIP and its crew ended up spending two additional months on Pyramid. Craig, Dawn, Karen and Shelby had been working together with the teams from Pyramid nearly seven days a week. Karen and Craig had taken a camping vacation for one week of that to one of the other continents on Pyramid. Dawn had spent that week reviewing the state of the mines and oil fields on Pyramid, their refineries, their factories, and their communications.

Karen said, “We need to get back to Pangaea. Without communications we have no idea what they are doing or if they even still exist. They could have been attacked and destroyed and we would never know it.”

“I know you are right, we need to leave right away.”

Dawn agreed, “We need to take what we have learned about this new fuel and get back there. Assuming they are still okay, which I hope, they may have made advances we need to know about to bring back here. Shelby and the Pyramid teams can keep working with the new fuel. Maybe Pangaea and Jill in particular can figure out some better ways to use this new solid chemical fuel.”

“I will miss you all, but I need to get back with Bill, my husband here. I have only been with him a half dozen times in the past six months.”

It was sudden, but they left Pyramid for Pangaea the next morning Pyramid time. Morning or afternoon meant little when leaving on a three month voyage.

# CHAPTER TWENTY-ONE

# ADVANCES ON PANGAEA

Three months later they arrived back on Pangaea. They approached slowly broadcasting their arrival on radio for fear that some patrol from Pangaea might identify them as an enemy ship.

Jill, Jane, Adriana, Sandy and Estera were there to meet them. Sandy hugged Craig, “I’ve missed you,”

Adriana reported, “My programming works for guiding a missile to a point in space to be preceded by a second by the beam arrival. In fact we used one of our missiles as a target and the beam zipped right through just before the nuclear warhead exploded at the point it would have been inside a larger ship. This was all videoed from one of our space fighters at a safe distance.”

Jill had been anxiously waiting her turn, “The main problem is that we calculate that we can only detect the enemy ships a few million miles out, and that is only visually. The only way we could do better would be to station ships out in space, but we would have to have space stations placed way out millions of miles and equipped with the beam and missiles. The stations would be sitting ducks once they fired, supposing the enemy did not find them first.”

Jane followed, “As Adriana said, we have tested the small beam weapon and are building larger beam weapons. As Jill said, the missiles are slower than light and any beam is limited by the speed of light. I have been studying your computer records. What caused the destruction of those continents on two different planets? It seems to me that they had a beam weapon. Maybe those two planets you found fired at each other with beam weapons across light years and destroyed each other. That would imply that they fired at each other and waited years for their beams to destroy each other."

Dawn said, "That still makes no sense. Maybe they were attacked by the enemy that shot at us. If so, that would mean that they have a beam weapon to destroy Impervium far superior to ours. If they have the weapon, they must also have a defense for it."

Jane replied, "But that would be a huge source of power in their ships and maybe they would not anticipate a tiny beam weapon like ours. Plus according to your surveys there was no background radiation indicating that they had no nuclear weapons."

Jill interjected, "What we need is some form of warning device out in space around the planet, but the enemy could be approaching faster than the speed of light."

"Or, we need to figure out where they came from using SHIP and spy on them." Craig replied. "As in the best defense is a good offense?"

"Maybe. You are the captain, but I thought we agreed that our first order of business is to set up a defense of our colonies."

"You are right. But, if we have some spare engineering time I would like to start work on our tracking system. Maybe we can sneak some time to go back to Earth and tag that spaceship watching the Library of Atlantis. Not only that, but I still want to get into that Library to see what is so important."

Dawn replied, "I think we all want that. We have some engineering time for some of the people now, so we can start working on it. With Shelby's new solid fuel, we could stand off further and just send in a missile. But remember, we don't want to see the missile before we use it defensively and give them a chance to develop a defense."

"I have a defensive idea. With this new fuel that Shelby developed, we have the potential to use a small missile and fly faster than the speed of light. If we could maybe put a space station out at a distance from Pangaea and pack in some faster than light missiles, Adriana could develop software to guide that missile back to us and give us more warning of approaching enemy ships."

Dawn said, "Great idea, Jill, let's do it, but a space station implies putting people and food and water out there in space. To be effective we need to put it out about a billion miles from the planet and that means a prohibitive supply line. That is the nearest Neptune and Earth ever come. By the speed of light that is still only four hours warning. With acceleration time a manned flight would take months to get there, too long for a manned station. The station needs to have artificial intelligence to see radar and visual clues and recognize a ship coming in and then fire off a message missile back to planet with that information."

Two months later they had developed a new unmanned spaceship to go out a billion miles at half light speed. These ships would be spaced out around their little solar system. Each sentry ship had radar, cameras, infrared sensors, mass sensors, and every sensor they could think of all linked into a computer program that would watch for any large item approaching their solar system anywhere near light speed. Upon identifying a potential target the sentry ship would launch a missile back to Pangaea that could arrive in half the speed of light or about seven hours. The missile could not be any faster because it had to accelerate from the sentry ship then decelerate to reenter Pangaea's atmosphere to deliver the data to humans and computers on planet for further analysis. In the even of a potential threat space fighters would be launched with the missiles and small beam weapons read to go.

# CHAPTER TWENTY-TWO

# RETURN TO PYRAMID

Karen, Dawn, and Craig left for the three month journey back to Pyramid. They carried with them all the detailed designs perfected on Pangaea. When they arrived it had been eight months since they had left Pyramid.

They were surprised when Shelby did not greet them. Craig presented the new developments on Pangaea and recommended the reformations of the research and development teams from the year before. During a break in the presentation Craig asked, "Why isn't Shelby here? Is something wrong with her?"

The planet president answered Craig, "No not at all. She just had a baby boy and both Shelby and her boy are doing fine. Her husband, Bill, is overjoyed and bragging to everyone. Her housemates are showering her with all kinds of attention. She will probably be back in research by the end of the month if she can tear herself away from her baby. "

"That's great."

"I have a question. Why does Shelby and most of the rest of your family seem so young? It has been nearly fifteen years since we left Earth and I would swear Shelby is no more than twenty-three. I know that she must be at least thirty-eight which is a little late to be having a first child. Since I have been seeing videos of you and your family since before the migration from Earth none of you seemed to have aged more than four or five years while those of us in the colony have aged fourteen or fifteen years, depending on when we joined the training for the colonies. A few engineers here were in their twenties when they met you and Dawn four years before we left Earth. They are now in their later thirties and you and Dawn are a little older, but are still in your twenties."

Not wanting to tell anyone about how SHIP worked to reconstruct their subatomic structure every time they teleported Craig answered, "Maybe it is all the traveling at faster than light speed. Einstein's theory of relativity stated that you might travel away from Earth at near the speed of light and when you returned you might not have aged while those on Earth did age."

"That IS what Einstein said, but could we go back to Earth now and arrive before it was destroyed."

"No. We have returned to Earth several times and it does not work that way. Maybe we should show some videos we took from a distance from Earth. As we traveled back to Earth we caught some images of Earth that, because of the speed of light were before the impact, during the impact, and after Earth was hit."

"I think that might be good to show on our television channels. We have some colonists pining for Earth and wishing they could go home. There are even some that think Earth was never destroyed and they were just told that to make them happy to live on a colony while Earth is still totally intact."

"Okay, I'll download the videos into your computers at the television station. I think you should advertise them for at least a week in advance of showing them. In fact, it might be good to have Dawn narrate them for the planet wide audience."

"Great! I think everyone on Pyramid would want to see them. Maybe we can shut everything down for a few hours to make sure everyone can watch."

Dawn did the presentation which was broadcast to everywhere in the built up areas of the planet of Pyramid. Keep in mind that this built up area was less than that of Iowa. Not everyone had television sets but most had tablets or personal computers that could show the one television station. While bringing sorrow to most, it was also closure. There was no question of going back to Earth for generations. Those that were in doubt about the truth of

Earth's fate were no longer in doubt. The conscientious was that their efforts to make the planet their own were redoubled. Pyramid would be their home for generations. What they did today would be for the benefit of their children and grandchildren.

For those working on the new weapons in any remote capacity were shown part two of the videos, the attack on SHIP. Of course SHIP did not appear in it but the video was obviously taken from their ship. This gave the teams the why they were working on planet defense and cemented the idea of a bad guy in space that might find them unprepared. This redoubled their efforts to develop these defenses.

Three weeks after SHIP's arrival at Pyramid, Shelby showed up. She was forty pounds heavier than she had been last year. "Guess what guys, I'm a mother. I have my own baby boy, Andrew. Bill thinks Andrew was three weeks premature, but we know different don't we? I could not help putting on weight. Bill doesn't care that I did, he is proud to be a father again with his young bride."

"I was kind of hoping I could teleport to SHIP and get my figure back."

Dawn said, "We want to see Andrew when we can. I would imagine Craig wants to most of all since he might be the father. I think you should go into a training program to lose weight. Going to SHIP to reverse changes is like cheating isn't it?"

Craig said, "I agree with Dawn. Yes, I want to see Andrew and am glad that you were able to have a healthy child, but an instant change to pre-baby condition, might make all the women want the same for themselves. We don't want to create jealousy and have a thousand women trying to find out how they can get teleported. Not only that, but there are probably some vain men too."

"Are you saying you don't want me anymore?"

“Shelby, shame on you. You know that is not true. You have his child. You are the one that wanted to stay here and knew that you were severing your ties with the SHIP’s crew.”

“I’m truly sorry. It’s these hormone imbalances. I have been pregnant for nine months and now I am lactating more than enough for one child. Before joining the SHIP I stayed in shape with a lot of exercise and cheerleading. There is no cheerleading with no teams, but there is plenty of work to do now that Andrew is born.”

“I shouldn’t have gotten so lazy and let myself gain the weight, but the other women in my family were so happy that I was going to have a child, they spoiled me. They insisted on doing all the work like I was fragile or something. I did my scientific work, but got no exercise and I was fed everything I liked to eat. Bill wasn’t any help either, he treated me like a china doll, getting me any treats he could find and not letting me lift a finger to do anything. The pregnancy and birth could not have been any easier.”

“I’m sorry too Shelby. Some of the engineers have already been asking how we have stayed appearing so young while everyone on planet shows their true age. I gave them a lie about Einstein’s theory of relativity and flying away from Earth and returning without aging while people on Earth aged. A lie, but easier to explain than SHIP’s teleportation that keeps our age constant once we have reached our prime. The longer we go the more questions there will be about our agelessness. If you lost forty pounds and regained your muscle tone overnight, we could not hide it.”

“I know, I know, Craig. I wanted to be a normal colonist living on a planet. It was selfish of me to even ask.”

“I have been working on the solid fuel I invented. I can easily make the fuel load any size you want and can stop and start the burning within limits. I chain together packets of the fuel and can have the burnt one reignite the next packet or not depending on whether we want continued propulsion."

Karen asked, "What do you think of sending out an unmanned sensored space station to detect the enemy at long range and then using faster than light missiles to send data logs back to the planet."

"Of course. We have actually tested some missiles that were faster than light, but we don't have the computer software to operate a remote space station or the sensor software or software to guide these fast missiles. We have the space fighters flying around Pangaea around three hundred thousand miles, too close to do much good, but they have the missiles loaded with nuclear warheads."

"We brought back software and the advances made by Pangaea. In particular, we have the perfected small beam weapon and the plans for a planet based high power weapon, but they were just starting to work on developing a space station and have not yet developed anything faster than light to bring back the data." Dawn explained.

Craig said, "We need to take your advances, get back to Pangaea, see how their software is going, and then get back here with the software. But before we go, let's go see that baby of yours, Andrew, and meet your family here on Pyramid."

"Bill, I would like you to meet Craig, Karen, and Dawn. They are all members of the original crew that planned the escape from Earth and found this planet of ours."

"Pleased to meet you Craig, Karen, Dawn."

"Karen, I have heard so much about you. You don't look like an Amazon warrior girl."

"I am an American Indian, one of the few left. I am not an Amazon warrior. I was heavily into to hunting and living off the land. Craig needed a girl that could serve as a protector that had hunting and camping experience to help protect his family when exploring new planets, if we ever found one. Craig was limited to finding a girl for his crew of six women and one man. Craig chose me as the protector and explorer. We found three that were suitable for humans before the escape. Dawn was the first of his family because of her interest in astronavigation. He knew that it was critical to be able to navigate space with no compass reference or even the reference to our sun."

"I can see why he chose you Dawn; you are very beautiful, but so young. I figured you must be pushing forty by now."

"Thank you. It has something to do with long term space travel maybe aging us in reverse. We really don't understand it."

"Well, whatever. I am glad that Shelby is so young. I am closing on forty and feel very lucky to have a wife that appears to be in her early twenties."

"Craig, I don't mean to give all the attention to your women, but I expected you to be some kind of god. I hope you will excuse me, but look pretty average. In fact compared to a lot of the male colonists, you are not what I expected to be leading this whole colonization of the stars."

Craig was pretty ordinary looking. The average male colonist was over six foot and under six four and very fit, usually with large muscles and a six pack. Craig was only five ten and while in good condition, was not muscled like an athlete. Because of what SHIP had given him, he run further and faster than any other human. He could lift weights in excess of the most powerful human power lifter. But, he did not look like he could."

"Pleased to meet you too. Shelby has been bragging about you and your family. Not least, she has been bragging about her new baby, Andrew."

"Yes, we are all proud of Andrew. I was a little worried about Shelby not having a first child until her late thirties, but she assured me that she was as fit as a twenty year old. Several of my wives have had children close in age to Andrew, but they had at least three children before their last one came along in their late thirties. One is on her sixth child. It seems like when Shelby got pregnant several of my wives got pregnant shortly thereafter so the house will soon be full of babies. Maybe her getting pregnant just inspired me to have more sex with the rest of my family. I had gotten wrapped up in my work and gotten complacent in keeping my women happy. It changed my desire for women. I had been feeling old and when Shelby got pregnant, it made me feel more virile."

Craig, Shelby, Dawn and Karen knew that Shelby was physically in her twenties even though technically she was in her later thirties. The SHIP's family was all in their twenties with some holding as young as twenty-one or whatever the genetic prime was. None of SHIP's family was over twenty-seven physically with the average being twenty-four. When they joined the crew they were between nineteen and twenty-one and that was fourteen years ago.

Each of Bill's women, except Shelby, made one of their culinary specialties. Bill even had a couple of bottles of the wine they were now producing on Pyramid. The dinner with Shelby's family was a huge success.

Two days latter Dawn, Karen, and Craig left for the three month high speed trip to Pangaea with the formula for the improved fuel developed by Shelby.

When they briefed the rest of Craig's team upon their arrival, Jill exclaimed, "That is marvelous. With a faster than light missile we could hit them without them even seeing it coming. Whether radar or visual, or any sensor, the signals can only come at the speed of light, therefore if we could hit them far enough away, they would never even see it coming. For missiles returning with data, we need to slow them down for reentry to the atmosphere and landing, but if our space fighters or a space station could detect them far enough out, our missiles would only be visible to any sensors when accelerating."

Then Dawn continued, "For them to attack us, they will have to slow to less than the speed of light before they get here, unless they have anti-gravity like SHIP, but we saw no sign of that when they attempted to attack us near Earth."

Jill added, "Just like when we were hunting for planets, SHIP could detect likely planets, but without slowing down to far slower speed we could not know if the planet was habitable. To really study a planet for life that was compatible for humans we actually had to orbit a few times. Then we teleported to the surface for further study. Of course they would just need to confirm where we built our cities. I would think they would need to get down to a few million miles. There are billions of stars with billions of planets. The enemy would have to have an approximate location to find one of our planets and then they would have to slow down to observe the planet. At only two light years out, the image would be two years behind the real planet they were looking at. I guess technically, since we have been here for fourteen years, they could observe our colony from fourteen light years out, while we could not detect them that far out. They could have already launched missiles at us years ago and they have just not arrived yet. They could have fired beam weapons years ago and we have not felt them yet. We don't even know what direction they would come from."

"My point that we need to find their home." Craig added. "That is really frightening to think that they could fire a beam weapon or missile years ago and the impact may still be coming. How could you defend against that?"

Jill said, "With this new fuel we could put a space station out there a billion miles like the orbit of Neptune in our old solar system. If we could have an unmanned space fighter with a beam weapon and a few missiles on board, we just might be able to stop them before they got close enough to identify which planet. Adriana where are you on the software?"

"I think it is about ready, but I would have to say it is at a beginning beta stage. There is a lot more testing, and some of that testing needs to be in space using Dawns astronavigation so the station will know where it is at and where the sensors are oriented."

Craig asked, "How long would it take to get a prototype space station out a few hundred thousand miles for testing purposes?"

"Maybe two months." Jill answered.

"And a month of testing using manned space fighters to test its sensor sensitivity." Adriana inserted. "How long to develop an unmanned space fighter with beam weapons and missiles."

"I assume that would be without nuclear warheads." Jill replied.

"Absolutely without nukes." Dawn ordered.

"Just a guess, but I would say we could build a test model of an unmanned space fighter in the same two months as the test space station."

Two months later they had an unmanned space station with two unmanned space fighters on station at four hundred thousand miles from Pangaea. They tested the sensors first with another unmanned space fighter that flew out to a million miles, turned around and attempted to fly past that station at one hundred thousand miles. It worked except that the beam weapons weren't armed to actually be used and the missiles were relatively slow. There were glitches but all thought they were minor and did not prevent building an actual space station.

It took six months to get a single space station on guard one billion miles from Pangaea. It had six space fighters docked to it with real nuclear warheads and armed beam weapons. The station would use its sensors at maximum sensitivity for the longest range and the software would sort the data to determine if anything the sensors saw was natural like a rock in space or an intelligently built ship. The station had fifty small message missiles that would quickly accelerate above the speed of light toward Pangaea then decelerate and land on Pangaea to bring back sensor data. For testing purposes, the first dozen confirmed sightings of intelligent ships would not be fired on, the thirteenth would be fired on.

The initial testing was with Craig and Karen in SHIP flying by without any stealth capabilities never getting within ten million miles. Rather than waste a message missile, the station transmitted its sensor data directly to SHIP. No problem, but after three more flybys they discovered the maximum range of the sensors was five hundred million miles. At that range the station could not tell the ship from a rock. When SHIP was traveling at just above the speed of light the sensor range was nearly zero, but space fighters could have been launched from the station to catch the intruder from behind up to twice the speed of light. Above the speed of light the beam weapon was useless as the space fighter was faster than the electronic beam that was limited to light speed.

To modify the software in space, SHIP approached the station at low speed, parked hull to hull and Jill teleported over with a back pack carrying some minimal clothing and all the tools needed by Adriana and Jane while on the space station. Adriana had to get undressed and teleport over with Craig. Adriana had been with SHIP for over a year, but had not teleported enough times for SHIP to teleport her any other way. Craig then teleported back to SHIP and brought Jane over the same way Adriana had to teleport. As she undressed for the trip, Jane commented, "When I get time I want to study this teleportation to understand it better. I understand what you said about not confusing me with my equipment in teleporting, but I do not understand how SHIP can learn the rest of you so well that you can teleport with whatever you want." Karen, Dawn, and even Sandy had come along to this point one billion miles from

Pangaea. They wanted to be together as a complete family if anything bad happened.

Before they teleported back to SHIP they armed all systems to defend to the best of the space stations ability. Adriana said, "I programmed SHIP into the station software so that the space station will allow free passage and also to recognize its own space fighters and missiles as friendly. Any other intelligent ship will be fired on with everything the station has. In the event all targets are destroyed any launched space fighters with remaining weapons and fuel will return to dock with the station and await another attack."

Dawn said, "We will need over several hundred space stations with thousands of nuclear warheads. We cannot rob all the nuclear material on all of the colony ships. We need to develop some uranium mines, refining and separation facilities to make our own plutonium for these additional stations and missiles. Just the space fighters would need huge amounts of plutonium just for the beam weapons. We just will not have enough even if we could strip all the plutonium out of the colony ships. We are talking years of work to build all this equipment and mine and refine the plutonium. While on this trip I have worked with SHIP and we estimate it will take a good ten years to defend Pangaea in all directions. It would almost seem that we should be working on a force field or something to protect the planet."

"I think we need to find their home planet or planets. If we could find that, we might be able to put space stations along their most likely route to us instead of trying to sensor a ball with a billion mile radius which is over twelve billion square miles or over sixteen billion square kilometers. You have said they could not detect a faster than light missile, so why not send one at the enemy ship near the Library of Atlantis?"

"Nice idea, Karen, but not so good." Jill answered, "A two or three pound object traveling at faster than the speed of light is likely to obliterate the spaceship on contact when all we want to do is mark it to leave a trail in order to follow it back to its home. To do that we have to decelerate the missile to maybe one hundred miles per hour before impact, which would mean firing retrorockets for an extended period of time which would definitely get their attention. Firing the retrorockets to slow down means blanking out our own sensors and then not having fuel to chase them when they escape."

Craig supported the offensive plan, "Maybe SHIP, with full stealth, could simply sneak up on the enemy ship and follow it home when a missile from nowhere came in at sub light speed? SHIP?"

SHIP replied, "Unknown. I have the capabilities to hide from any Earth sensors, and any Martian sensors that existed when I was built, but I have no idea about the enemy capability. I do know they had weapons beyond any Martian weapons which are far beyond earthly weapons. They had weapons not in my memory when they fired at me. I think I could adapt my defenses against those new ones they fired at us."

"SHIP, could you pull within range and fire a tracking missile at them?"

"Two problems. I am strictly defensive. I could equip your small shuttle ship you used many years ago with a tracking missile of your design, but my programming prevents me from overt offensive actions. You gave me the other; once a missile comes at them they may be able to adapt their defenses also which might make all your missiles known and therefore not able to do their job."

# CHAPTER TWENTY-THREE

# BATTLE FOR PYRAMID

Craig and family had spent over a year on Pangaea working on the new space station. Pangaea had tested, perfected, and then deployed one space station in orbit one billion miles from the planet and over one billion miles from their sun. It orbited their sun like a distant planet only fifty feet wide and camouflaged to look like a rock. The second station had been launched and was also working like a champ. It would take a minimum of ten years to surround their sun with space stations for defense and years to develop a uranium mining and refining industry to produce the plutonium needed to power the stations and provide for warheads while still keeping most of their colony ship power for their colonies.

In addition, they would have to start producing new Impervium for all of their space stations, space fighters, and missiles. This was the one thing they knew all about from producing the original colony ships for the escape from Earth, but they did not have the supporting industry they had on Earth. It could be twenty years before they consider building another colony ship if they needed one, but within three years they could be building space fighters and missiles from scratch instead of stripping a colony ship using Jane's low power beam weapon for peaceful purposes.

Craig said, "We need to take what we have learned about space stations, production of plutonium, and production of additional Impervium and get it back to Pyramid. Who is coming along and who is staying here?"

Jill volunteered first, "As the main engineer, I need to stay here and help direct the effort. I would like to take the vacation with you, but I am afraid to go off and leave things undone here."

"Okay, Jill. I understand. Estera, I presume you are staying here."

"Right. This is my home now. I will stay with my family, have more children and grow old with them. Physically I am eight years the youngest of the adults thanks to SHIP."

"Dawn?"

"I am torn. I am the organizer, the trainer, but I might be more help on Pyramid than here. Besides, I have always been with you. I am going."

"Karen?"

"Of course I am going. I am not a scientist. I am the adventurer. Of course I am going."

"Sandy?"

"I have been here on Pangaea for three years. I have studied the animals here ad nauseam and all our Earth animals are reproducing on their own. I have done experiments with cross breeding the buffalo type cattle here with Earth domestic cattle, and they are not really compatible. I am more than ready to go with you Craig. I have missed you. I didn't really realize how much until our time together here on Pangaea. I may not leave again."

"Okay Sandy, I have missed you too. Jane?"

"I think I should go with you to see how the beam weapons for Pyramid are going and help with their space station development.

"Adriana?"

"You have all my software, to date, loaded into SHIP ready to help Pyramid. Everything seems to work as planned. But maybe going to Pyramid will fire some new ideas. Besides, this was my first time away from Pyramid. I would like to consult with some of the other computer people I used to work with doing research on Pyramid. The space station, the crewless space fighters, the missiles, even the aiming and firing of all the beam weapons is dependent upon my software. I have tested, fixed, retested, and fixed the software here until everything seems to be near perfection, but Pyramid is a whole other story. I need to be there. I would also like to just see my old friends on Pyramid, besides, Craig, I want my share of you."

Okay then. Estera, I am glad you are content here. Jill, you have been away from SHIP for a couple of years now. I miss you, but I also appreciate you dedication to the effort. Yes, you will probably be more valuable here. We are not trying to invent or engineer anything new on Pyramid, just implementing things developed here."

"Keep in mind that we might be gone a long time on this trip to Pyramid. They are far behind Pangaea in their defense. Depending on how things go, we might just possibly go on to Earth again. I am still concerned that we need to learn more about the enemy. While Pangaea is way ahead of Pyramid in defenses and new industry, it is still many years from having a good defense. I have always believed that a good offense is the best defense. That is why the United States was the world leader on Earth."

Jill asked, "What if you decide to go after the Library of Atlantis again? Won't you need six women?"

"True, I don't think we will be attempting that or chasing after the enemy on this trip. However, I suspect it may be a couple of years before our return here. I wish there was some form of communication between colonies. It has been five years since we even checked on Vulcan. We should go there just to check on them, make sure they aren't developing offensive weapons, make sure the enemy hasn't wiped them out, and make sure there hasn't been any natural catastrophe. That will add to our time away from here. Jill, are you prepared to be careful and alone for two or three years?"

“Estera is here. I have all my engineers and mechanics and plenty of work. Are you trying to talk me out of staying?”

“Sort of. We will return.”

“Well, I will take care. This planet may not be developed, but it is more civilized than Earth ever was.”

So, coming along to Pyramid were Craig, Dawn, Karen, Jane, Adriana, and Sandy. Jill and Estera were staying behind. The voyage would be three months and they planned on at least six months of work, likely up to two years before returning to Pangaea.

Sandy monopolized Craig for the first week of the three month trip. “I really cannot get enough. It has been five years on Pangaea. Now that I am back on SHIP I am back to twenty-one years old and haven’t had sex on SHIP in five years.”

After a week, Sandy decided to give Craig more of a break as she started watching videos that SHIP had recorded in the five years Sandy had stayed on Pangaea.

Adriana took advantage of Sandy’s giving Craig some time and pushed Craig into the bedroom. Okay, Craig, we had sex when you first brought me onboard SHIP. I thought you were taking advantage when I had to be nude to teleport while the women who had been with you longer did not have to, but now I know you were not just taking advantage of me. I have seen you take the other women and Sandy just spent almost every waking moment for the past week here in your bedroom with you. I think it is high time, I took your cloths off and we see how we fit in your bed.” She was not wearing much and it quickly dropped to the floor. Then she walked up to Craig, “Hands off. I will do it.” And then undid each of his buttons one at a time. When his shirt was off she moved up chest to chest and while looking up at his face removed the rest of his clothing.

An hour later they were both temporarily spent. “Now I understand why Sandy wants to spend so much time in your bedroom and why the rest of your women take turns. I had sex with my husband on Pyramid before my accident, but I had only been with him a few months and had to share with five other wives. It was a chore and expected of me as all women and men were expected to have sex to have as many children as possible to populate the colony, but this was amazing. I was a virgin when the colony arrived on Pyramid and sex was not fun. When we arrived, it was almost mechanical to select a family. There were attachments made during the trip here, but many of us made no attachments. I was busy with the computers from the time we left Earth. Be prepared to give me my share from now on. Are you always this good?"

"Well, I am physically twenty five. I have the sex drive of a nineteen year old, and we all are kept in perfect shape on SHIP. I enjoy it differently depending on who I am with. Sometimes it is athletic, sometimes slow and tender."

"How do I compare?"

"Unfair question. Dawn and I have been together for nearly twenty years now. The rest of the original family has been with me for sixteen years. Estera and I only had sex maybe once a month if that. It was a matter of being together for years. Shelby was maybe once every two weeks and was just to satisfy her occasional need to belong. Jill was Dawn's roommate in college and I have known her just as long as I have known Dawn, but it was a couple of years later that Dawn convinced me we needed more scientists to succeed and SHIP dictated they had to be women. Jill was a natural. She was a tease before joining the family and very athletic in bed. Karen was the last of the original family members. She is very demanding, but only once a week unless it is just the two of us exploring and then it might be twice or three times in a twenty four hour period. Sandy and I had an animal attraction from the being. If anyone loves me it would be Sandy. If it is possible, I love all of my women in different ways. Except for you and Jane, we have been together on this mission for years in very small quarters on SHIP. I enjoyed sex with you very much. I cannot say I love you like I know that Sandy, Dawn, or Karen do, but this is the first time we have had non-mandatory sex and the first time we have talked about personal subjects.

Three months later as they approached Pyramid they saw some anomalies. They were still seven million miles out when SHIP said, “There has been a problem on Pyramid.”

“What kind?” asked Dawn.

Sandy was already at the monitor since she had been away from SHIP for nearly five years. “There are blackened spots in the city like there was a major fire, but aren’t the houses all solid stone?”

Adriana said, “Yes, our population has still not filled all of the houses in the central city of Pyramid.”

Always true to her defense of the family, Karen asked, “SHIP! Are there any enemy ships around Pyramid?”

“No. Neither do I detect space fighters in orbit.”

Craig asked, “Is there still life? Television or radio signals?”

“Yes.”

“Put their television on the main screen in the living area where we can all see it.”

The broadcast from the one station on Pyramid was panning across the blackened area. “Citizen’s of Pyramid, we won this first battle. There were an estimated twenty enemy ships that came in and started firing weapons at a half million miles. Our space fighters were hit first, but when we launched the second wave of fighters we hit them hard. Our beam weapons and nuclear tipped missiles caused five of them to just explode and another six to immediately turn away apparently damaged. The other nine enemy ships went into full retreat. Shelby, our defense minister, will take over and give you her assessment."

Shelby filled the screen in front of a map of the inhabited part of Pyramid. After destroying our space born fighters they started bombarding the city of Pyramid. Our first wave of fighters was caught by surprise, but when we had our second wave take off, the enemy underestimated our weapons which worked as advertised. The beam weapons burned holes in their hulls and the small missiles just flew inside the hulls and exploded their nuclear warheads destroying five of the enemy ships. The remainder of the enemy fleet tried to immediately flee our solar system and we managed to severely damage another six. The remaining nine managed to escape. This shows we need to put more effort into defense because there is a real threat they will be back and this time our beam weapons and nuclear tipped small missiles will not be a surprise. We need all able bodies to work their regular jobs to maintain our civilization and spend at least four hours a day as unpaid overtime working for our defense industry. This applies to mothers with young children too. The weakest member of each family should act as baby sitter for the other five women taking care of their children."

"Fortunately, the enemy did not have time to study our city and did not hit any populated areas, only vacant houses we have not yet used. We are searching for any survivors that might have been wandering in these vacant areas looking for either new housing, office or factory space. I will give you daily reports for at least the next week."

Karen and Craig teleported to the surface of Pyramid as soon as they arrived in orbit while the rest of the family stayed on board SHIP. SHIP was in full stealth mode while searching the heavens for other ships.

Craig waited for Shelby outside the television studio. Because the population was watching the television show and since SHIP was in full stealth mode, no one had seen Karen and Craig appear. When Craig saw Shelby exiting he walked to intercept her, "Shelby, we just arrived and heard your broadcast during our approach. I guess we got here a day too late."

"Oh, Craig and you too Karen, you surprised me. Yes, yesterday we lost nearly half of our space fighters. We had so little warning. Our space fighters were only a few hundred thousand miles out when the enemy ships started firing from a million miles. Naturally the first targets were everything we had in orbit. From ground radar and telescopes we saw them moving in closer. They started firing at ground targets in the city as we launched our second set of space fighters. The space fighters started firing at the enemy ships as soon as they left the atmosphere. They didn't act like they knew what hit them and immediately went into full retreat. Well, you heard my television report."

"Yes. Do you think they detected the beams or the missiles?

"There was no indication that they detected anything. I think some of their ships blew up and several more were hit during full retreat. I wish we had gotten all of them, but our space fighters do not have much range because the human crew has to be able to get back home in a few days."

"We have the fix for that. I brought everyone with me this time except for Estera that is a permanent resident of Pangaea and Jill that felt she needed to stay there to push for more production. Between Jill, Jane, and Adriana primarily we have developed and tested a totally automated space station that can sit out there for years at one billion miles out with its own bevy of automated space fighters that have beam weapons and missiles. Thanks to your new solid fuel, we also have faster than light message missiles on the station to send data back to Pangaea in the event the station detects enemy ships."

"If we had that technology we could have destroyed all the enemy ships without any losses except to maybe some robotic space fighters and maybe the station."

Karen jumped in, "What Craig has not told you is that to effectively maintain a billion mile shield around Pangaea it will take more Impervium and plutonium than we can take from the colony ships. We have to maintain some of the colony ships for computer centers, providing electrical power, hospitals, and emergency shelters. As you know, Pangaea did not have prepared stone buildings and roads like Pyramid does. Most of the original buildings were built of wood although newer ones are concrete, and because we had to start mining from the start, there is not enough metal for metal reinforcing of the concrete."

Craig interrupted, "Which brings me back to my idea from years ago. We need to track the enemy back to where they are based. If we knew where they are coming from, we could concentrate our defenses in that direction rather than three hundred sixty degrees a billion miles out."

Shelby exclaimed, "But I think we were able to tag a couple of the escaping enemy ships. A derivative of my solid fuel can burn very slowly and should leave a trace amount of some rare gas that I think could be tracked by SHIP."

Karen excitedly said, "Craig, we need to quickly offload our designs and software to Pyramid and get SHIP tracking the enemy. By leaving soon we should be able to keep our distance from them and track them. If some of their ships are damaged and since they lost the battle, they should be headed home. If we can track them that far, we will know where they came from. With that data, we could put some stations near their planet and have super high speed missiles to warn about them coming our way."

"Right you are, Karen. Building a billion mile ball of defensive stations would take many years and one heck a lot of Impervium and plutonium for nuclear power for the stations and warheads for missiles. We could watch their planet and watch for an armada leaving their planet and have a larger automated missile that can bring back that message at many times the speed of light. These messenger missiles will not be limited to some small undetectable size but can be large enough to carry lots of solid fuel and even a nuclear engine like the colony ships. The ships will be dramatically smaller than the colony ships because they will be carrying just a computer chip with warning data versus a whole colony with food, supplies, and equipment. It probably would not be much larger than four or five times the size of the manned space fighters here on Pyramid."

Karen added, "Probably one of our space stations would be enough to monitor one planet. If we can find their base we could maybe get one station there in two or three years without having to produce addition Impervium or plutonium."

Craig said, "I would vote for at least two stations, one nearer and one further. If they find the first one, they may not think to look for another further out. If the first one fails for whatever reason a messenger missile will be launched. Since Pyramid is behind Pangaea I recommend that you produce the automated space fighters first and simultaneously start working on at least two long range space stations with the long range messenger missiles."

Shelby said, "Consider it done."

"By the way, minister of defense?"

"Yes, Karen. Because of my history with the family, and because of my new fuel and understanding of the defense systems, I did not really have a choice."

"I still have not lost the weight, but I did have another child that was definitely Bill's. I have been too busy to worry about my weight. The other members of my family were doing most of the child rearing of my children. Because I was leading the development efforts here on Pyramid, I decided to bring my children to a child care center near my work so I could see them during breaks."

Dawn asked, "Why are you crying Shelby? That sounds like a good way to see your children and work too."

"Because we decided to build the nuclear processing in a remote unpopulated area of the city. It was hit during the attack. I had gone to headquarters for a meeting and left my children at the processing center. They died instantly and I am once again without children."

"Oh Shelby, I am so sorry." Dawn said.

"No, it was my fault. I had the choice of leaving my children with my family or bringing them to daycare at my workplace. I made the mistake of pride. I thought my being around them more would be good for them. It was because I was too possessive. I did not have to take on the Minister of Defense job. I should have left it to someone more political and just kept to my chemistry work. I could have done most of the work from home, but again, the power went to my head. I just want to get away from all this."

Craig asked, "What about Bill and your family here?"

"Bill asked me to leave. He and his other wives never wanted me to take his children to a daycare. It hurt their pride. They took it that I did not trust them with my children. Bill tried to talk me out of taking the job of Secretary of Defense. I should have listened. Now I have no family at all, and that is why I wanted to stay on the planet, to have a normal family life. I screwed that up royally."

Craig asked, "What would happen if you resigned and came with us?"

"I don't fit anywhere now. I am between generations here. Because SHIP kept me young the colonists aged fourteen years while I was not aging. Their children are twelve years younger than my physical age. At the same time, I have been on Pyramid for six years now and am getting old and fat and don't fit with you anymore."

# CHAPTER TWENTY-FOUR

# THE CHASE FOR THE LIBRARY

Dawn said, "Craig, don't you think that SHIP would teleport her up as the same age and health as when she left?"

"Yes."

"Shelby, we are one person short right now. If we cannot track this enemy fleet back to their base we may want to try again for the Library of Atlantis and we will need the formula of six women and one man again. I think SHIP will put you back the way you were before you decided to stay on Pyramid. If you decided again to stay on a planet, you could try Pangaea. Estera is quite happy there. Jill doesn't plan on staying there forever, but she stayed behind to continue the development there."

"Yeah, I don't want to stay here on Pyramid after all that's happened. Once we get the planet defenses up there is no need for me and I think I have done all the chemistry there is to do here. I never thought being in my twenties was a bad thing, but when the whole planet is either in their late-thirties or children, I am out of place. I think I will be content to stay with SHIP. I do love all you guys and I appreciate you a lot more than before. Can you forgive me for killing your only child?"

"Shelby. You never told me you had gotten pregnant with my child."

"That is what I wanted, but I failed. I killed your child."

"Don't say that. Of course I would have liked to know I was a father. I know you were very happy to have a child and that you had another child shortly thereafter that had to be Bill's. Did you ever tell him it was mine?"

"No. Having a child that he thought was his really ramped him up and he not only got me pregnant, but all of his other wives. Losing that child because of my jealousy putting them in childcare at my work alienated Bill's whole family against me. I was too involved in my work here and ended up isolating myself and my children from the love and safety of being with the family. There were only twenty children killed in the childcare center and that included both of mine. One of which was also yours."

"Shelby. Cut that out. It was not your fault. You were new to that family and I can understand why you wanted to keep the children nearby. There was no way you could know the attack would come so fast. It was pure bad luck that they happened to destroy the childcare center while they were there. They were most likely aiming for the nuclear processing center there and missed."

Dawn and the rest of the women took over reassuring Shelby of her innocence.

SHIP was able to teleport Shelby, but when she arrived on SHIP she had lost years and weight and looked exactly as she had when she first decided to live on Pyramid. Shelby said, "Wow. Talk about recovering from having a baby. I am just like I was a few years ago."

They downloaded all of their information and software developed on Pangaea and took off after the trail of the enemy ships. It was easy to detect the unusual chemical signature in empty space left in the wake of the ships. SHIP could not trace it at high speed, but still traveled at multiples of light speed. Shelby had left her job in the hands of her second and made one last visit to her Pyramid family explaining that she was rejoining Craig's crew.

After the first week of travel Dawn announced, "They seem to be headed right for Earth's solar system. SHIP, display their course on the hologram display in the living area so everyone can see it in relation to Pyramid and Earth."

SHIP did as directed and said, "I agree with Dawn's assumption. My calculations make it nine-five percent probability they are headed for Earth's solar system."

Craig responded, "Maybe we should accelerate and slow down just outside of Pluto to see if the track is correct?"

Shelby warned, "The tracer won't stay forever. If we're wrong we will lose the track completely."

Dawn said, "It should take about two months at high speed to get to our old solar system. If we don't find the trace there, we would use another two months getting back to this point in space. What would four months do to the trace?"

"I don't really know. This is purely experimental. I do not know what the dispersion would be in the vacancy of space, but we know there are plenty of rocks moving around out here that could throw off everything. The more time goes by the more likely we will lose our ability to trace."

Karen suggested, "Why don't we try a series of jumps? If we go fast for a week and find nothing, we will be back to this point in only two weeks instead of four months. That would allow us to get wherever faster while reducing the risk of losing the trail. Shelby?"

"Let's try a few really short jumps. Let's make it half a week jump and back here in a total of one week. See if it works?"

It took them thee months instead of four months, but the jumps cut down the time from six months at the speed they were going to just less than three months. The trail did not go to any planet in the Sol System, but went right on between planets. Unfortunately the proximity of the planets and the asteroid belt caused them to lose the track."

"Shelby, where did the track go?"

Dawn answered for her, “Well Craig, I think they flew right through our old solar system. Our old solar system is unusual as we have found out in our travels. Most stars do not have solid planets that are big enough to support intelligent life. The sol system has several in ever expanding orbits. We have found asteroid belts around other stars, but not in common with several planets. In addition, many of the planets have large moons orbiting them. Now add in some gas giants and the gravitational pulls destroyed our trail.”

“SHIP, pull out every stealth trick you have as we get into the solar system. Girls, shall we try for the Library of Atlantis again?”

“Okay, Craig, but SHIP, also put out every sensor you have and look for any intelligent interference as we approach Earth. I don’t think any of us want to be surprised again.”

Karen added, “The enemy ship fired on us last time and did some minor damage, but they may have added some weaponry since then. If there is the slightest indication of another intelligent ship anywhere, get all of us out of the solar system as fast as possible.”

Dawn directed, “SHIP, I want you to be constantly calculating a clear path out of the solar system for a quick getaway.”

This time SHIP sent down a preprogrammed robot with video, infrared, sonar, and special software developed by Adriana over the past years since their previous attempt. This would give the robot the capability to explore the underground library system without touching anything but the floor and then after exploring every isle and room would return to its starting point. SHIP, in turn, would teleport the robot to that previously identified open spot and then on a regular space of time would attempt to teleport the robot back on board.

As during their previous attempt, SHIP could not detect even the location of the robot after it was transported, let alone see what it was seeing. The library obviously had some sort of shielding.

While waiting, Craig asked, “Why did the enemy ship know where the library was and then try to destroy us when we just started to explore?”

Karen replied, “I think that they had some evidence of where it was, but could not get in. I wonder if it is in an Impervium shell that the enemy could not cut through. Maybe Jane’s beam weapon is that unique. We never found anything else that would harm it, which is why we named it ‘Impervium’”.

Dawn picked up, “I think you may be right. They must not be able to teleport like SHIP does. We don’t know how SHIP does it and SHIP even claims it does not know how it does what it does. We have come to the conclusion that SHIP is from another intelligent race that is far ahead of the enemy and may be ahead of the Martians. If the enemy knew where it was but could not cut through the Impervium and could not teleport through it, they were still unable to get inside. They must think that the information in there would be worth placing a guard ship on the library for years. They were not there when we evacuated from Earth, but sometime during the first fourteen years since we evacuated, they did place a guard.”

Karen continued, “I think maybe they saw our evacuation and obviously observed the destruction of our civilization, so came and placed a guard on the library in case someone came back to explore it. Maybe they think we got our technology to escape Earth from the library and they wanted that information for themselves. They don’t know that we did not even know its location until long after we were gone.”

When Karen paused, Craig continued the thought, “That makes sense. They may have thought that we were regularly tapping the knowledge there during our development phase. I think you were absolutely correct in your analysis. They apparently had the same information about its location that we found. Maybe they did not know where Earth was until we made our escape and then the destruction. Maybe they just want to cut off our accessing it again. Maybe they have the information, but don't want us to have the knowledge stored there."

Karen added, "Then, I hope SHIP can detect them before they can fire on us again."

Sandy got into the conversation, "Okay, so we know we can send in a sensor package and teleport it back. I think Adriana's software will make our robot work correctly, but we have no idea just how big the library is. It may be hours or even days before it gets back to its starting point. Then, if we don't get attacked before then, we don't know if SHIP can teleport a living being into and out of the library."

Karen said, "I will go alone to test that. We know the air is okay there. If SHIP cannot get me back, you will find that out and not try again with humans. Craig can go pick up Jill from Pangaea and still have his formula of six women."

"Karen, I don't want to risk you."

"Craig, you have no choice. You wanted to come here to get into the library and we all agreed to give it a try. SHIP needs you. I don't want to die or get left there, but I am more expendable than you are. I am the family defender. This is a threat to the family we have to test out. Sandy is our doctor and zoologist. Adriana is critical to software that we need more and more of. Dawn is your alter ego and her opinions are needed. Her knowledge of astronavigation exceeds that of even SHIP. Jane has proved herself crucial in the beam weapon that might be what saves humanity and already saved Pyramid with only minor damage. Sorry, Shelby, I know it was a major loss for you, but the planet was saved. Shelby has developed new fuels and weapons. I am not a scientist, I signed on for the adventure, challenges, and risk. I have to take that risk for the family."

"Okay, Karen. I can't argue your logic. I wish SHIP could teleport a live rat from Earth and teleport it there."

"So, I am your human rat. And Craig, we have teleported together when we had to watch each others back and saved each other several times, but this is not an animal or plant, this is some form of shield around the library. If one can go and come back with no harm, then anyone can. If I cannot then no one can. It is a yes or no, there is no mutual protection required."

"Okay, but let's make sure we don't have an enemy attack; let's get the robot back; and then we have to analyze what the robot finds, before we even try to teleport you."

Dawn suggested, "Before we send Karen down to the library alone, let's have SHIP serve up a raw steak, teleport it down, wait a few minutes and teleport it back up to see if it can teleport the closest thing to Karen we can."

"Are you comparing me to steak?"

"Yes, I guess I am. Got a better idea? It might just save your life."

"Danger! I am accelerating out of the solar system. I detected the enemy ship coming from Saturn."

"How did they detect us?" Craig asked.

"I had to drop part of my shields to teleport the robot to the library for just a second."

"Is there any pursuit?"

"No. I do not believe their ship can see me in full stealth mode."

"Why did no one on Earth detect you when teleporting me back and forth for years?"

"It was only a mass reading. Earth did not know what to think about reading a mass in geostationary orbit that only appeared for a fraction of a second and at different places in orbit. Most times there was no detection. When there was a blip on their sensors, they considered it a flaw in their equipment or some kind of interference. Apparently, the enemy ships are more sensitive or know more what they are looking for. I was invisible visibly because I was blending with the background by changing my exterior color from as seen from any angle. I absorb radar and sonar as just a source of energy. However, I cannot hide my mass when teleporting and my mass is like a very large asteroid or small moon."

"Okay. How about launching one of our tracer missiles at the ship when it stops and then we turn off your stealth until they give chase and then track them back?"

"I cannot fire an offensive weapon."

"It is defensive."

"Cannot do. My programming will not allow that. I could let you fly the lander with a missile, but you would be detected before you could get close."

Dawn suggested, "Let's go look at Saturn and its moon Titan. Maybe they have a temporary base there with the sensors that have found us."

"Okay. Unless anyone has an objection we will maintain our distance from the solar system until we can come toward Saturn on the side away from the sun, so we maybe see their base without them seeing us."

SHIP spent the day moving into position and then started its approach. Nothing was found as they approach Saturn and its major moon, Titan.

Dawn said, "Saturn and Jupiter are gas giants, how about their base being in the asteroid belt?"

SHIP replied, “Probable. Do you wish me to search the asteroid belt?”

“Yes. Start searching in order of size: Ceres, Pallas, Vesta, Hygiea. The asteroid belt is closer to Earth and the Sun. The Earth for convenience of spying on Earth and nearer the sun for energy. The bigger they are the more gravity. Ceres is about five hundred fifty miles in diameter with Pallas and Vesta a little over three hundred miles. If they do not have artificial or anti-gravity the more mass the better to live on for months or years. Mars is too near Earth and their base would be easy to find.”

At two million miles SHIP found their base. “Their base is on Ceres. I will display on the screen in the living area while I maintain position.”

“There is a dome that is emitting heat. Estimated interior temperature is seventy-five degrees. There are three exterior tunnels, one to a ship like the one that attacked me and one to an empty pad. That pad probably belongs to the ship that is now near Earth. The third pad is larger and may be for resupply from their home world. I have found no other intelligent structures. It appears to be an outpost of the enemy. Why they chose an asteroid instead of Mars that is much more hospitable or even Earth where even humans could survive again, I do not know.”

Karen added, “They are probably there to hide from other intelligent creatures that would automatically head toward Earth and still close enough to watch Earth if they have anywhere near the incredible optics of SHIP. Mars is nearest to Earth as far as being benign and therefore likely for us to look at closer. We did not expect them to come from the asteroid belt. They probably approached Earth from the direction of Saturn just to throw us off.”

The ever cautious Dawn said, “SHIP, let’s stay at this distance and survey the other large asteroids. There may be more than one base. We need to make sure that we know the enemy as best we can before taking any action."

"I want to mark their ships somehow with that tracking missile developed by Shelby. Both of them. I want the tracking material to stay intact until they leave the solar system and then start leaving the trail again. If we can coax them to head for home, we can then track them. We don't want the tracking material to be just wasted polluting the solar system and making it hard to track them. Shelby, can you do that?"

"No, but I'll bet with Dawn's knowledge of astronavigation and Adriana's software, we can make a couple of them here on SHIP with material produced by SHIP. Then we could maybe fire a tracking missile at the one on Ceres right now and one at the one near Earth whenever we can get up close."

"SHIP?"

"Craig, I can make the missiles to your specification from my nuclear structure, but I cannot fire missiles."

"Okay. I will take the shuttle and fire the missiles from it and mark the ship currently just sitting there on Ceres, then return to the stealth of SHIP. No, Karen, I can fly the shuttle and I am the only one experienced at it. I do not anticipate any defensive action from Ceres. SHIP says all their sensors are trained on Earth and are not even looking elsewhere."

It went perfectly. The small missile attached itself to the top of the enemy ship at the opposite end from the dome where it would not be seen without a hull inspection. The inspection was unlikely as the enemy did not know it had been attacked and there was no air on Ceres meaning an inspector would have to be fully suited for extraterrestrial work.

After Craig was back in SHIP he asked, "Okay, that went perfectly. Assuming the tracking is set right to start when leaving the solar system and assuming that we can induce them to head for their home world and assuming they do not inspect their hull, we were successful. Now, what is the next step?"

Karen volunteered, "I think we now go after the ship near Earth and try the same thing. I think SHIP should go within range of the missile, you take out the shuttle again, fire the missile and immediately get back in SHIP, we retreat, and then we decide on step three."

"I agree, girls? Does everyone agree or have other ideas?"

Sandy disagreed, "Craig, why do you have to take that risk? Maybe they won't attack our colonies again."

"I am pretty sure they will attack us again, but this time they will be more prepared and bring an overwhelming force. Since they know where Pyramid is, they may be assembling an armada right now. We may already be late."

"Okay, Craig. Do it, but stay within feet of SHIP so you can get back on board immediately and then we run."

Dawn supported Sandy, "I agree with Sandy. They have a mass sensor and are looking for us; they will spot the shuttle immediately. Their attack on you will be at the speed of light, and the shuttle will not go that fast so you will only have minutes to get sheltered by SHIP and then SHIP has to immediately retreat faster than their weapons can fire at the point in space where we were."

It took a few hours to get into position. Again the tracking missile went perfectly, attaching itself to the enemy ship. The enemy ship did not seem to detect the missile, but did detect the shuttle when it emerged from SHIP and fired beam weapons and missiles. Craig fired almost immediately when he cleared SHIP and then immediately retreated back inside SHIP which was gone before any weapons arrived at their previous location in space. Beam weapons can only travel at the speed of light, and SHIP was faster than any of their missiles which went out into space self detonating when there was nothing for them to hit.

They had already determined their next step. SHIP could not fire weapons, but could push a small asteroid on a course determined by Dawn. The relative speed was only a thousand miles per hour, but the course of the asteroid would intersect the orbit of Ceres and impact the base on Ceres within hours. SHIP was safely away waiting. The enemy detected the rogue asteroid an hour before impact. The ship at the base lifted off with ten minutes to spare. The rogue asteroid was only ten feet in diameter and only traveling at a speed of a thousand miles per hour, but it totally obliterated all traces of the enemy base. The enemy ship stopped in space to watch the destruction of their base and SHIP detected their sensors searching space for a cause of the small asteroid hitting their base. SHIP also detected a tight signal sent to the other enemy ship near Earth. After a few hours both enemy ships were on an intersecting course toward the side of the solar system furthest from Pyramid. SHIP stayed back tracking their course visually, by radar, and by infrared. The two ships were near each other on a parallel course when they left the solar system, presumably for their home world. As soon as they got to the Ort Cloud at the far reaches of Sol's gravity, the missiles attached to the enemy ships started leaving a chemical trail that SHIP could follow. SHIP dropped back to be two light days behind the enemy ships so as to remain undetected by the enemy ships.

As expected the enemy ships accelerated slowly apparently not having antigravity like SHIP had so acceleration was like the colony ships, limited to what the inhabitants could withstand. It was quite leisurely for SHIP to keep up. When they got to the halfway point to their destination, the enemy ships would have to decelerate like their acceleration, slow.

Craig asked Dawn, "Do you think they are heading for their home world?"

"That was our plan, right? When their temporary base was destroyed due to the unlucky collision with the rogue asteroid, they could not stay as they would have limited onboard supplies. They probably had enough on board each ship for weeks or maybe a few months, but they may be months from their home world with their leisurely acceleration and deceleration. If they had tried waiting without a base of supply they could end up starving out in empty space without their home world knowing what had happened to them."

"I think you are right. Are there any guesses as to where they are headed? A known planet? A known star?"

"Too early to tell."

Sandy asked, "Now that they are not protecting the Library of Atlantis, could we go back and explore and then catch up with them later?"

Karen jumped in, "I think Sandy has a good idea. We could take a quick look and easily catch up with those two ships. In fact, we could spend a few days there and still catch them before they achieve light speed. It's no wonder Pyramid's weapons inflicted damage on their fleet. Missiles can be built to withstand fifty or more gees of acceleration, but their ships appear to be limited to what the inhabitants can withstand."

# CHAPTER TWENTY-FIVE

# EXPLORING THE LIBRARY

Dawn, Adriana, Jane, and Shelby all agreed so Craig had SHIP break off the trail and return to Earth. SHIP teleported what should have been the robot from its preprogrammed waiting area and successfully brought the robot back on board. They spent two days just studying the robots recordings. The robot had spent four days just roaming around recording images and creating a map of the library. The videos showed banks of what had to be computers. It had wandered into a closed door with the standard seven buttons that could be pressed by six women and one man as they had found elsewhere the Martians had been and wanted to secure something. All corridors seemed to lead to this interior room at the center of the library. Some of corridors had what appeared to be stacks of stone tablets, others paper products, but many were banks of electronics of some sort. Wherever the robot had wandered lights had come on as it entered a corridor and went off when it left so the library was still powered.

Dawn remarked, "Imagine a system still powered after thousands or maybe tens of thousands of years. Maybe everything still works. The air is standard Earth, but slightly on the dry side, not enough to spark electronics but probably to preserve other knowledge stored there."

The next step was teleporting a steak to the location of the robot recovery, waiting an hour, then teleporting it back to SHIP and testing it and having SHIP test it.

"Okay my turn."

"We are only going to leave you there for five seconds and then back here to discuss so look quick but do not move even one step, in fact try not to move anything but your eyes."

"Okay, Craig. You sound like my mother."

"Okay, Karen. Are you okay?"

"Just fine, Craig, but I did not even have time to blink."

"What did you see?"

"A row of shelves with tons of boxes and stacks of what must have been books but the printing did not appear to be either Martian or any Earth language I am familiar with."

"Ready for a second trip?"

"Sure."

"Try to stay in the same position. Do not move your feet as SHIP is teleporting a location, not you per se until you are free of the shielding of the library. In one minute, try to get your arms, hands, and head back into the same position as when you first arrived. At one minute and ten seconds SHIP will teleport you back."

Karen returned, "Why don't we just plan to jump near what appears to be the control room?"

"Should have thought of that. Ship, teleport the robot directly to the area near the control room door and order it to look as many directions as possible, move over five feet to the left of the door and teleport it back in one minute."

"Done, Craig. Returning it in five, four, three, two, one" and it was there. They all reviewed the sensor data for thirty minutes.

"Okay SHIP, teleport a two pound t-bone steak to that location five feet to the left of the door. Leave it there for ten minutes and bring it back." They waited until it returned, Sandy inspected it, and then SHIP analyzed it.

"My turn." Karen declared.

"Just stand there and inspect the door, don't move. In one minutes SHIP will teleport you back."

One minute later, Karen said, “It was just like the robot showed. Beside the door there are seven large palm sized buttons. Should open the door. I saw no windows. I want to do one more test. I will go down, use chalk to mark the placement of my feet, walk over to the door, try pressing one of the female buttons, then walk back and put my feet inside the chalk marks. Okay?”

Craig replied, “Okay, I’ll give you two minutes. Get back in the chalk marks in one hundred seconds. That will give you twenty seconds of leeway.”

Two minutes later Karen appeared back on SHIP, “Like the other buttons, when you press one, nothing happens. The door appears well sealed. You can hardly see the seam around the door.”

“Before we all go down, I want to be teleported with Karen to verify that SHIP can teleport two of use. We will both chalk mark our arrival and get back to the marks before the time is up. Two minutes should work. Okay?” Dawn said.

Two minutes later, both Dawn and Karen reappeared on SHIP with no ill effects. “What about Adriana and Jane?” asked Shelby?

“We will need everyone to open the control room if that is what it is. Karen, why don’t you take Adriana’s and my clothing? I will bring Adriana down. Then Dawn, you go next and take Jane’s clothing. Then after I get there, Sandy and Shelby can teleport down. Everyone will chalk mark an outline of their shoes. Adriana and Jane will have to outline their feet. That way we can get back to our original locations before you teleport us back. We’ll reverse the sequence on the way up with Shelby first and Karen last. SHIP, are you sure that there are no enemy ships around?”

“I detect nothing that does not belong in the solar system.”

“Okay, we will stay for no more than twenty minutes. We will be back on our chalk outlines before the twenty minutes are up.

SHIP, start teleporting us one at a time in exactly twenty minutes. I want to make sure we are back on SHIP in case an enemy shows up while we are down there."

When they were all down and Adriana and Jane dressed except for shoes, they went to the buttons with Craig in the middle as all of the locking buttons as before with three girls on each side and on a count of three they all pressed their respective buttons. The control room door slid into the wall. Karen entered the room first to make sure they would not be trapped in the control room. She found a single button to close the door and pressed it. The door closed and then reopened when she pressed it again. Then they all entered the control room. As Jane entered as the last of the seven, a voice was heard in what appeared to be a language. Then the language changed again, and again and again. Finally, Karen said, "That was Spanish. Hola."

The voice came fast in Spanish. Karen said, "Favor de hablar inglés?"

"Welcome. You are now in the control room of a repository of Rogulatory. I was built on a planet in a star system called Carina. My planet is not visible from Earth since it is about 8,000 light-years away from this planet you know as Earth. We pushed into space because our dual stars were predicted to go Nova and eventually did. This repository has been located on Earth for one million Earth years. This installation was placed here to monitor intelligent life forms. Rogulatory has been a space capable race for two million years. In exploring this galaxy, these repositories were placed on every planet found that had the conditions necessary for development of intelligent life."

"We listened to your electronic communications as soon as they were invented. We have monitored conditions on this planet you call Earth since the repository was placed here. As soon as the conditions were right for intelligent life, a space ship was placed in orbit to circle the planet and attempt to bring the intelligent life to the point that they were able to find this repository and enter this control room. Since you are here, you have the intelligence to wisely use the knowledge here. At some point, maybe our races will meet in peace. You call yourself humans. Humans evolved from animals that had to fight for their food and have used your emerging intellect to invent technology to kill each other."

"You call the ship, SHIP, as if it were intelligent, but in reality it is only a computer, just like I am only a computer. SHIP, as you call it has been selecting captains and giving them new technology for many years. Actually, the planet you called Mars gained advanced knowledge quicker than Earth did. When we detected intelligent transmissions from Mars, SHIP was moved to orbit Mars and helped that civilization first. When the Martians evacuated for Earth SHIP provided them with the warning of the ending of their atmosphere and provided the technology to move to Earth just like SHIP did for you Earthlings before your evacuation from here."

"When the Martians came here, SHIP worked with both Martians and Earthmen to achieve new levels of technology. With the Martians it was the development of faster than light space ships and with Earthmen of the time; it was making copper and brass tools. Whereas the Martians were peaceful, the Earthman made swords and shields. The Martians could easily have cleaned the Earthmen off the Earth, but they had concentrated their population on a continent doomed for extinction, requiring them to leave Earth before the inevitable confrontation with Earthmen."

Dawn asked, "Does the planets we call Pyramid and Pangaea have repositories."

"I am not familiar with those names. I would need to know their locations based on my formulas for navigation of this galaxy."

Craig said, "Our time is running out, we need to get back to SHIP."

SHIP told Craig by telepathy, "No, the shielding is now open to me. I have located you and made contact with the repository. According to the repository, I can teleport you out directly from there and next time teleport you directly into the control room."

"SHIP contacted me and says we can teleport back and forth from SHIP directly to and from here whenever we want now that we have gained access. SHIP is in contact directly with the repository."

Dawn said, "So we can stay here."

"No, I am going to tell SHIP to start teleporting us back to SHIP one at a time. Shelby, you were last, so you will be first."

Craig telepathically told SHIP, "Start bringing us back now." Shelby disappeared and within two minutes they were all back onboard SHIP.

"SHIP, how much information can you or have you gotten from the repository?" Craig asked.

"I now have total access to all electronic records in the repository."

Karen asked, "Does that mean the enemy ships might come back, chase us away, and they will have access to the repository?"

SHIP answered, "No, when SHIP leaves the security goes back into place like before."

"That's good. Sorry, I interrupted Craig. Would you ask SHIP if it has any weapons that we could use against the enemy ships?"

SHIP replied, "The repository has scientific data that will advance our knowledge, but was not programmed with offensive weapons."

Karen argued, "But didn't you give us the atomic bomb?"

“No, I gave humanity the atomic energy secrets that was to be used for peaceful purposes to provide power like you used for your colony ships.”

“Why didn’t you give us the secret for tapping the core of the Earth for generating power and the power pyramids used by the Martians?”

“I gave the Martians the atomic energy first and then the power pyramids and the high efficiency electric propulsion units to power their vehicles. The power pyramids are used to increase and direct the natural static electricity in the air. Humans misused the atomic energy secrets and were not ready for the geothermal energy and power transmission. In fact, Tesla was given the secret for power transmission, but tried to rush it into production. He skipped over the capability for extremely efficient electronics. Everything back then was vacuum tubes that required a lot of electricity compared to later transistors and then the Martian power transistors. Therefore Tesla needed very high power transmissions far in excess of what the Martians needed. What you know as the Tunguska event was an enormously powerful explosion that occurred in June 1908 was really an attempt by Tesla to transmit electricity from New York to Paris. He used too much power and missed. The power was to be bounced off the ionosphere like the United States military HARP experiment and be reflected down to Paris. Instead the power beam arrived at Tunguska in Siberia. Besides, humans had no way to use that kind of electricity.”

Dawn asked, “So SHIP, have you been holding out information?”

“No. I have had many captains that overstepped their capability. When I recognized that, I removed them and selected a new captain. Tesla appeared to be one of my best. He did not use his knowledge to create weapons. He used everything for the betterment of humanity, but he got too far ahead of the technology of his time. Therefore, I never gave out the secrets of power transmission to any other captain. Craig, my current captain, has used his knowledge wisely and followed all my rules and taken my calculations or advice."

Adriana said, "Let's go back down that we know we can teleport at anytime."

Karen argued, "We need to track the enemy and see where they came from."

Craig said, "I agree. SHIP, can you pull the data on the enemy and then get after the enemy trail?"

SHIP replied, "I pulled all enemy data from the repository and am now following the trail." The crew could see Earth, then the sun, then the solar system receding. The enemy may be who pushed the celestial body into Mars thousands of years ago. The Martians thought that it would have missed Mars all those years ago except that it got a push from the enemy and destroyed Mars. Then just a few years ago that same body that was coming for Earth and it arrived early. I think that was an attempt to catch humans before they were able to escape into space."

Karen said, "So who are the enemy?"

"Martians. There was a civil war on Mars with a small number of Martians that tried to use technology to take on the rest of the planet. Martians were generally peaceful, but they did not want to be ruled and the technologists left Mars and went to some other planet. The Martians just outnumbered them too much and were not that far behind in technology. The ones you call the enemy were originally Martians but have been separated in space for at least fifty thousand years. The Martians that came to Earth were taller than earthmen were because of the lower gravity on Mars. After fifty thousand years of living on a different planet they could look different than Martians living on other planets."

"If you say Martians were peaceful why did some of them try to take over?"

"According to Martian history, they were an isolated group that thought they knew better than the common Martian and tried to rule them by threat of high technology weapons. The Martian citizens overwhelmed them with horrendous losses with only a few thousand of the technologists escaping the planet. When they came back in a few years to try again, Mars had defensive satellites in place that were able to repulse them. When the Martians discovered the interstellar body from the Ort Cloud coming for them there were suspicions that the technologists were behind it, but with no way to stop it, they evacuated as many as they could to Earth. Because Earth was larger with a heavier atmosphere the Martians took several generations to assimilate to Earth conditions."

"Some of the strongest Martians were able to move out of Atlantis and ruled in places like Egypt and Greece where they were accepted like gods because of their supreme knowledge of engineering, healing, farming, and all other sciences."

"Eventually, these colonies in Egypt thought they should be independent of Atlantis. The story was told by Plato from information in the Library of Alexandria. Atlantis sent out a force of ten thousand troops that were easily conquering hundreds of thousands of earthmen when they discovered a small asteroid headed directly for Atlantis. The troops were recalled from Africa and Europe back to Atlantis where they once again headed into space looking for a more peaceful home. Eventually, the rulers of Egypt and Greece from Atlantis died out from interbreeding and Earth society fell backwards. This is where SHIP started educating Earthmen like Aristotle."

"The repository recorded everything."

Craig asked, "How is the repository powered and how does it know what is going on everywhere on Mars and then Earth. The repository is powered by geothermal energy converted to electricity. There are sensors planted all over Earth to record everything from the movement of people and animals and even voices within ten miles. If exposed to any digging or light the sensors self destruct to dust, but with thousands of sensors only particular areas are lost. Of course not everything could be recorded that way, but enough to keep track of general trends and popular ideas. Of course all electronic emissions were tracked in detail."

Craig then asked, "What was repository going to do with all of this information?"

"Store it until accessed by someone with the right combination of women to a single man with enough technology to begin to understand. It also required the proximity of SHIP. Many captains of SHIP overflew this location, but they were alone. Craig is the first captain that figured out that SHIP would only allow one man and filled out his crew with the six women required. Therefore, once the control room was unlocked with SHIP overhead, everything was in place and the electronic shield dropped."

"What about your original manufacturers? How are you going to get the information to them? Who and where are they?" Craig asked.

SHIP replied, "You have no name for where they came from. They are the same as you but different, but that was hundreds of thousands of years ago. They may look very different due to the planet or planets they now live on, if they still exist. The repository opens up with information to anyone with six women and one man with a ship that has technology similar or more advanced than me. I was a critical key in case some other family managed to gain access to this repository. If someone managed to breech the repository shell with the right number of six women and one man but did not have a spaceship nearby like me, the repository would self destruct."

"You said there were repositories all around the galaxy, have any others been destroyed or accessed?"

“I have no knowledge of that. I only know the locations of repositories planned before my construction. Once a year, the repository sends out a data stream into space reporting any new data in the last year. Because it can only travel at light speed, I do not know if anyone received it or had the capability of decoding it. If it reaches my makers, it could take hundreds or thousands of years to reach them since I have no knowledge of where they settled.”

“Did the repository know where the enemy ships come from?”

“Only that they leave the solar system in the direction of Alpha Centauri.”

# TWENTY-SIX

# RETRIBUTION

"Okay, SHIP, let's try to follow their trail while we still can. We can come back to the repository later."

SHIP answered Craig, "I downloaded some of the pertinent data I thought you would want based on the questions you asked and research you've done. I am now leaving the solar system and can accelerate and still track the ships."

"What did you download?"

"The power pyramids and the static electricity receivers that powered the automobiles and aircraft of the Martians. I also have information on capturing dark matter and dark energy. Some Earth scientists thought there was more dark material than visible. I only have the data on how to capture it for space travel beyond our galaxy."

For three months they continued following the chemical trail from the two enemy ships. They did not want to catch them as they wanted their home planet so they had to go relatively slow for SHIP. Whenever the chemical trail got too strong they knew they were getting too close and slowed. They pasted several stars without habitable planets. As they traveled, Dawn, Jane, and Shelby studied the data gathered from the repository. Karen, Sandy, and Adriana took turns with Craig when he was not studying the repository results. Dawn, Jane, and to a lesser extent Shelby worked with the data until they were too exhausted to have sex with Craig. Craig, having been energized by SHIP years ago, could easily keep up with three oversexed young women.

After years of being cooped up on SHIP they had exhausted the movie and television recordings stored by SHIP. The library contained every trace of entertainment created by mankind in its history, but eventually, they were all watching reruns. You can only read novels so many hours a day. Everyone was kept in perfect physical condition by SHIP, but physical activity was limited to virtual reality and sex, so sex was a popular sport. SHIP had provided a multi-purpose weight machine, a treadmill, and a stationary bicycle all of which were linked to virtual reality, but there was no space for conventional sports to burn the energy of the physically fit young bodies. Craig still needed sleep each twenty-four hour period, but he never tired. Most of his sleep was a couple of hours of quiet time with whatever girl had visited him last.

Finally after three weeks, the enemy ships started slowing. They were like the colony ships, they had no anti-gravity so they had to accelerate for half the distance at the maximum sustained gravity they crew members could stand and then decelerate at the same gravity until they reached their destination. The enemy ships had been enroute for six months. SHIP brought up data on the enemy planet.

"It is slightly smaller than Mars which is much smaller than Earth, but has gravity near Earth's. Its star is older than Sol, but the planet is closer providing a climate similar to Earth. I can detect power pyramids around the planet. There are two hundred enemy ships in orbit and a number of small shuttle craft going back and forth to the planet surface. Apparently, their spaceships are not able to land and takeoff from the surface."

Dawn interrupted SHIP, "Doesn't that show they don't have Impervium and therefore their spaceships are too heavy to lift off and land on the planet"

"Yes, that is my analysis."

Dawn asked, "Tell me about their solar system."

"There are only two planets. One is gaseous and then the home planet. There are a number of rocks in orbit that were captured by their sun at some point."

Karen asked, "Are there some four or five miles in diameter near the orbit of their planet?"

"Yes, there are at least one hundred near their orbit. There are five in the exact same orbit, but millions of miles away. There are some only a few hundred thousand miles out that orbit at a slightly lower speed, which would make their year longer than the planet in question."

Karen asked, "Craig, why don't we do to them what they may have done to Mars and Earth?"

"Which was?"

"Change the orbit of one of those larger rocks to impact with their planet and take care of them once and for all."

Sandy objected, "But that would kill everyone on the planet."

Karen asked SHIP, "Estimated population and how many are on the enemy ships in orbit?"

"I cannot be exact, maybe five hundred million on the planet and twenty thousand in orbit. I do not see any ships ready to leave so they are lightly manned. I do not see any large ships like colony ships."

"So maybe ten thousand escape if we destroy their planet?"

"They would have some warning and their ships that are already in orbit. They could maybe get one hundred thousand on their ships before their planet was destroyed. Unlike Mars and Earth, they have hundreds of defensive satellites in orbit around the planet. They are not ships, so I did not tell you about them earlier, but they could easily repulse an invasion of ships."

Karen asked, "So, we need to send a rock into them too big for their defenses. Sandy, I know you don't like my idea, but they killed billions on Mars, then Earth, using the same thing."

Sandy argued, "We don't know that. Maybe they did that to destroy Mars after they were run off, but the rock that hit Earth was the same rock thousands of years later."

Dawn entered the argument, "Yes, but the Martians and even SHIP calculated that Earth had several more years before the rock came back toward Earth and what are the odds that it would even hit Earth after destroying the civilization on Mars? I think it had their help."

Shelby agreed with Dawn and Karen, "I think we should hit them as hard as we can. They tried to shoot down SHIP when we first found the Library of Atlantis and they attacked Pyramid when we had done nothing to them except run when they attacked SHIP. If they were behind the rock that hit Earth, they killed billions with only a few thousand escaping. As SHIP says, they already have the ships in orbit and could easily evacuate a larger percent of their population than we evacuated from Earth."

Jane simply said, "I vote for the attack on them if we are taking votes."

Adriana voted, "Let's do it. Craig?"

"I understand Sandy and I will have to admit that it is unlikely that everyone on their planet is our enemy, but just like Germany and Japan during World War Two, we have to stop them. I think it is very likely they destroyed two peaceful planets. Without any provocation on our part their planet was behind the attack on SHIP then they attacked Pyramid. All our work in recent years has been to defend against them and destroy their ships. The crews on the ships we were going to destroy probably had some peaceful members that had no choice but to do the job their planet sent them to do. We have to take them out while we can."

Sandy capitulated, "Okay. I will make it unanimous. We have wasted years on Pyramid and Pangaea developing defenses against them. If during an attack, we destroyed their ships we would still be killing innocent crew members."

Shelby added, "Our colonies are too small and our defenses too weak to stop their large fleet if they came in force. We have no weapons to take out their planet defenses and their ships. A large rock would do it."

Sandy asked, "SHIP, what is the largest rock in near orbit that you could direct into their planet?"

"There is one about three hundred miles in diameter that is comprised largely of iron that I should be able to decelerate from its orbit outside their planet. It is the size of a small moon and being made largely of iron, hopefully they will not be able to stop it. Once it loses enough speed their star will pull it the rest of the way because of the stars gravity. I will need help from Dawn to decide how long and how hard to push on that rock to make it swoop into their planet."

Dawn and SHIP with help from Adriana calculated the force and time needed to cause the rock to decelerate out of its current orbit and then be attracted by their star's gravity to circle into the enemy planet.

SHIP approached the rock in full stealth mode to not be sensed by the planet and gently docked with the rock. Then after three days of pushing with SHIP's antigravity drive slowed the rocks orbit by one hundred thousand miles per hour. Earth's orbit speed is approximately five hundred fifty-nine thousand miles per hour. The enemy planet was nearer their stars so its orbit speed was only four hundred twenty-two miles per hour. Between the deceleration caused by their stars gravity, the rock would intersect with the planets path in about one month with a speed differential of nearly one hundred thousand miles per hour. Assuming that the enemy could not appreciably reduce the rocks mass by firing at it, it could just slow the enemy planet enough to drop out of its orbit and into its sun in a year or two. The initial impact should destroy all life on the planet even more so than the rock that hit Earth.

SHIP backed off from the enemy solar system to watch. The family did not have much to do except study the data that SHIP had gotten from the repository on Earth and have sex. At other times they got together and played card and board games. Sometimes, they would play some of the old Earth computer games. If it had to do with quick reactions and shooting weapons Karen had the high score, followed by Jane, then Craig. If it was strategy, Craig was first, challenged by Adriana, followed by Dawn.

Sandy concentrated on data on the enemy planet, desperately wanting more data on the plant and animal life on the planet that was about to get destroyed. Shelby worked a lot with Sandy trying to figure out if any chemical differences on the enemy planet made them warlike.

“Sandy, they use chemical rockets to get their shuttles up to their spaceships. That is not high tech. I thought they were supposed to be technocrats.”

“Well Shelby, I would say that they have the pyramid beam technology, but maybe it just isn’t enough to achieve orbit. I have talked with Jane about the pyramid power and she says most habitable planets have an ionosphere that protects the inhabitants from their stars radiation. Jane thinks that the ionosphere layer cuts off the broadcast electricity from their pyramids. Look and see if they don’t use broadcast power to get to high altitude before they kick in their chemical rockets.”

“Sandy, you are right. Right now they are sending up five shuttles per day. They accelerate like aircraft until they reach five thousand miles per hour then zoom up toward space, kicking in their rockets to make it the rest of the way. When the shuttles land they use retrorockets to decelerate and then fly like airplanes back to their bases."

"I think that thousands of years ago, they may have been technocrats compared to the average Martian, but where the Martians kept advancing, they were into survival."

"I agree, Sandy, and to further complicate their advances, they concentrated on weapons of war and degrading the peaceful applications of that technology. It is hard to believe that they are still trying to kill people after a hundred thousand years."

"It may have been something in their genetics. A bunch of bullies inbreeding for thousands of years. I still think that there must be some percentage that doesn't even know what that their planet is still warlike or they are afraid to go against the others. There have to be farmers and workers that are not warlike."

"That may be, Sandy, but they have attacked humans twice for sure and may be responsible for killing billions on Earth. If they could guide an asteroid into Mars, then Earth, they had to have spent trillions in the equivalent of United States dollars just to use their technology to impact the orbit of an asteroid or a rock in Sol's Ort Cloud. It may be that the factory workers and farmers are so brainwashed after thousands of years that aggressiveness is now inborn."

They waited for a month before the enemy planet spotted the rock intersecting with their orbit. They launched fifty of their spaceships toward the rock and started firing their weapons on it. They put craters in it, but after a week it was still largely intact. They exploded nuclear bombs on its near surface trying to deflect it, but it just kept coming. Finally, a week before impact the shuttle traffic to the spaceships in orbit was constant as they tried to evacuate as many as they could. Less than twenty-four hours from impact their spaceships started leaving orbit in random directions. Some only went into an extended geosynchronous orbit.

Craig exclaimed, “Whoa! I did not expect that.”

Dawn said, “Calculations say that it was still one hundred miles in diameter when it hit. Debris was ejected out to fifty thousand miles destroying all of their ships in geosynchronous orbit.”

Karen said, “It looks like volcanic eruptions are going off all over the planet.”

SHIP said, “Their planet just lost twenty thousand miles per hour which should cause its orbit to decay into their sun in about ten years. It also changed the spin of the planet that could further erode its orbit. It may just go into a lower orbit nearer their star. Considering their star is dying, that may not be a bad thing in the long run.”

They watched the planet for a week. No spaceships returned.

Karen, the pragmatist, said, “How many ships got away and an assessment of the number of survivors.”

SHIP replied, “Twenty ships got away completely. These were not colony ships, and only had about one hundred crew members each for a total of two thousand survivors give or take a few. I do not really know how many were on each ship.”

Sandy sadly said, “Nothing living could have survived that and the dust cloud envelopes the planet.”

Shelby added, “Not only that, but SHIP detects that it lost about fifty percent of its atmosphere and most of its ionosphere.”

Dawn, the optimist said, “Keep in mind that Earth is surviving. It would have killed most of the animals including humans on Earth, but plants have a way of reseeding themselves.”

Karen asked, “Should we go after the spaceships that got away?”

Shelby replied, “We do not have any way to track them. They may not be to light speed yet, but we have no idea where they headed.”

Craig said, “I’ll bet some of them head to Earth which is now habitable again even though fifty percent is covered by ice. We don’t know what course they will take. They may know of other habitable planets that they will head to. If they go to Earth it will take them a minimum of six weeks from the speed of the ships we tracked here. If this was their only planet they don’t have the strength to take on one of our planets. I say we head back to Pangaea. We have been away for six months and it will take us at least five months to get back there.”

“They all agreed and SHIP went to maximum safe speed on the course set by Dawn for the shortest route.”

Five months later, Craig’s family was getting bored. They had spent most of the last year on board SHIP. They had watched all of the various television series recorded by SHIP and all of the movies that had been created by humans until they had exhausted the ones they were interested in. Only Jane was still working on shows she had not seen as a citizen of China before joining the family. Adriana, as a newer crew member still found a few she had not seen so Adriana and Jane became closer friends.

Dawn, Shelby, Sandy and Karen spent their time studying the new information that SHIP had garnered from the misnamed Library of Atlantis and having sex with Craig. All of them played various games and read books on their tablets. Dawn spent time just staring out into space. Karen spent more time than the others playing virtual reality games of hand to hand combat and firing weapons at virtual monsters.

Jane, after having sex with Craig, questioned, “Craig, I will be eternally grateful for rescuing me from that horrid planet, Vulcan. I enjoy having sex with you. It is better than I could imagine. I like the freedom all of you have, but I have what you would call cabin fever. I am wanting two things. I want to see the women on Vulcan again to see how they are doing and I am curious that they still exist. The enemy could have attacked them first and easily destroyed the colony there. Vulcan did not have the advantage of the new technology we developed on Pangaea and Pyramid. The second thing is that I would like to try spending more time on Pangaea and experiencing the freedom there. While there are not typical Chinese there, you did have a number of colonists that had Chinese ancestry. I would like to spend an extended visit with them. As an international gymnast from China I was able to travel around the world, but we were tightly controlled. Even talking with gymnasts from other countries was discouraged."

"I think both are good ideas, but Vulcan is months from Pangaea. We are planning to spend some extended time on Pangaea to get new technologies started and see how the colony is going. It may be months before we can leave for Vulcan and I would want to stop by Pyramid on the way. With our fastest communication being by SHIP from place to place we cannot know that any of our colonies have not gotten into trouble. I am sorry that Vulcan is at the bottom of my interest in the three colonies."

"But you could take me back to Vulcan sometime? I would not want to stay. I would only need a few hours there to talk with the other women. I fully understand your lack of concern for Vulcan, but you would be willing to take me there one more time?"

"Yes, but it may be a year or more."

"That is fine. I would also like to explore Pyramid some. I feel like I have grown up in a cave and just now seeing daylight. I never really saw Earth, only what my government wanted me to see. I do want time on planet before being closed up on SHIP again."

"We will have communications regardless of where you go on Pangaea. I see no reason you should not go exploring. I think everyone has cabin fever. I know Karen desperately wants to get out and maybe do some hunting and exploring the outback."

“Do you think she would take me with her on a short trip?”

“I’m sure she would. Everyone, including me is looking for some time away from SHIP. However, Karen will spend a great deal of time alone. We will all have radio communicators linking us to SHIP and each other.”

As they traveled closer, Adriana told Craig privately, “I would like to check in with Pyramid as soon as we can. However, I suspect I will be busy doing software to go with our new technology we got from the library. But I do worry about their safety.”

“I would not worry too much, they repulsed an attack by the enemy once and the enemy is definitely down to their last ships. I don’t think they would risk losing again with no hope of reinforcement. I think it is more likely they will go to Earth. That concerns me as we know there were some human survivors. However, there were so few enemy survivors that the two races coming together on an almost empty planet are not that great.”

“How many planets are inhabitable?”

“In twenty thirteen I read an article estimating thirty-eight billion planets in our galaxy the right distance from their sun. A number were thought to have liquid water making it likely.”

“But wouldn’t that mean that there may be millions of different intelligent species just in our galaxy?”

“You would think so, but so far all we know about is Mars and Earth. Even the enemy turned out to be from Mars. However, we have found five previously unknown habitable planets just on our own. Pangaea, Pyramid, Vulcan, and the two planets that had been inhabited until some disaster struck.”

# CHAPTER TWENTY-SEVEN

# USING THE LIBRARY DATA

When they teleported down to Pangaea there was a reunion with Jill and Estera. Estera now had three children. Jill had not had sex in over a year. However, when Dawn started downloading the new library data from SHIP to the colony ship computers Jill had been using, Jill was enthralled. "Wow! If we can develop these technologies, it will solve so many problems. Think of the technology of SHIP and imagine what you did not get from the library."

Estera asked, "Do you think we are safe from the enemy ships that attacked Pyramid?"

"We destroyed their planet even worse than Earth was destroyed by pushing a large two hundred mile rock out of its orbit into their planet. Only a few ships escaped and they were not colony ships. I think they will be more interested in just surviving instead of attacking. Besides, they apparently knew about Pyramid, but with billions of planets in the galaxy I do not think they know about Pangaea."

"Then, Craig, we could start concentrating on the new technologies?"

"I think so."

Jill exclaimed, "Dark matter, dark energy. Just think, this means potential fuel between galaxies. Even SHIP must replenish energy by absorbing matter and energy. SHIP has no problem traveling within the galaxy as long as we stop near a planet occasionally. Our colony ships might hold the basic design for scooping up dark energy and matter, but without antigravity acceleration and deceleration it takes too long to consider crossing the galaxy let alone going to another galaxy. You might have to accelerate for seventy years then decelerate for another seventy years. Simply impractical."

"So a new colony ship fueled by dark matter and energy, a true interstellar ship with unlimited fuel and speed. Between galaxies, it should be vacant enough to have incredible speed. Maybe a multi-generation trip?"

Estera interrupted, "What about the power pyramids? Imagine being able to harness static electricity. All of our electronics can be made incredibly energy efficient. If we can get a pyramid going and this technology going, we will no longer need petroleum except for lubrication, plastics, and clothing. We can eliminate nuclear mining, refining, and pollution."

Jill said, "I agree with the idea of no longer burning petroleum, but I still think we will need nuclear to just provide the basic energy for using dark matter and energy. I don't think gathering the dark will be free. I think we will still need nuclear."

Dawn interjected, "I am not totally sold on the power pyramids. Don't forget we found two planets that had been wiped out even though they had the power pyramids and I don't think it was the enemy. I have been thinking about it more and wonder if they did not overextend the power for some reason and wiped themselves out. I want to go back to those two planets and study them further."

"Maybe we could replace ship with a larger ship with exercise facilities like a basketball court, tennis court, running track. We know how to make Impervium for the ship and use dark matter and energy for fuel. It's not that I don't like sex with you Craig, but I cannot imagine being locked up on our little ship for years without landfall for some real exercise."

Jill said, "I'm afraid we can't do that anytime in the near future. I suppose if we built a new colony ship just for the seven of us we would have lots of space. We could have a forest."

Sandy said, "We could stock the forest with animals and birds that we might want to plant on other planets we find."

Shelby said, "That brings up another use for petroleum fuel to make chemical solid fuel rockets for shuttles. We still do not have the secret of teleportation. Every time we came to a new planet to visit we would have to shuttle down and back up."

Karen said, "Maybe we could build a ship big enough to hold SHIP for intergalactic travel. That way we could have the space to have everything."

Dawn answered, "Maybe someday, but our colonies cannot afford the effort to build such a ship today and not tomorrow. There is too much for us to do just keeping the colonies producing and multiplying. It will be thousands of years before we outgrow our current planets. Our colonies were chosen for their ability to support a colony for many many years with a young sun, good resources. Now that we have gotten rid of the threat of invasion from the enemy Martians we can concentrate on developing what we have."

Jill said, "I agree with Dawn, I'm a little cautious about the power pyramids. The high efficiency electronics are something we can use right now. But I still keep thinking about those two destroyed planets, and then there was Atlantis that was destroyed. Maybe tapping the heat of the inner planet is too dangerous. I think we need to go slow on that. However, we now have the knowledge to use petroleum and coal to produce electricity and the new high efficiency electronics will allow us to not have to run wires everywhere we want power. Exploring our planets will not be restricted to hauling fuel along with us and limiting our explorations. Interpolating what we have seen one power pyramid will allow us to explore a continent. The single continent of Pangaea might need three because of its size, but Pyramid would only need one for the main continent. Again, petroleum will still be needed to go to the other continents on

Pyramid. The one destroyed planet appeared to have one pyramid for each continent."

The next six months were spent on perfecting the high efficiency electronics and building a static electricity generator powered by petroleum instead of a geothermal power pyramid. Their generator was located near the oil fields they had already started developing.

Karen did a lot of exploring of Pangaea alone. She took Adriana with her once and Jane another time. Craig joined her in some of her explorations when she was getting too far from civilization for mutual protection.

Craig spent most of his time in the central city working and living with Jill, Dawn, and Shelby who stayed there with Adriana spending most of her time there and occasionally visiting the other cities, the petroleum fields, the mining and manufacturing community. Jane spent a lot of time visiting with the people of Pangaea, but spending most nights in the central city.

Estera spent nights with her family and she was pregnant again with her forth child. She could not teleport to SHIP while pregnant without losing the child.

They all teleported back and forth to SHIP frequently to consult the advanced computing of SHIP and to review plans holographically on SHIP displays in the living area rather than rely on more crude presentations on planet. The frequent teleportation to SHIP kept them all supremely healthy and after the first four months SHIP declared that Jane and Adriana could teleport fully clothed with equipment instead of naked with Craig. Craig said, "I will miss that part, but it will be more convenient for everyone. I enjoyed holding you two for the trip. However, that also means that I do not have to stay so close to be available for your trips to and from SHIP." Now the whole crew was able to independently teleport to and from SHIP.

Occasionally, Karen and Craig would have SHIP take them to other areas of Pangaea far from the colony. SHIP would have to stay in direct line of sight of the two in case they got in trouble. By line of sight, that was geostationary orbit line of sight. During those periods, the rest of the family had to be careful they did not get injured in some way since SHIP was out of range.

Karen asked, “Craig. In all my wanderings I have never seen a reptile on Pangaea. Only the relatives of buffalo, cats and rats. There are no bears, wolves, or other mammals.”

“You’re right. This is a strange planet. Almost the garden of Eden, but without the snake to mislead us.”

“This continent is roughly the size of all the continents on Earth before they broke up. The poles are covered with ice, there is either desert or jungle near the equator. Except for the animal life, it is like Earth millennia ago. There are enough resources here for many thousands of years of human life. Add in the Earth animals we brought with us as eggs and embryo and humanity should be able to survive here for a long time. Now when we add in the new technology that earthmen never had and it will be an Eden.”

They had both learned the dangers of the large cats and managed to avoid confrontations with them or scare them off before there was a problem. They spent their nights on SHIP as long as SHIP was line of sight. When they were within walking distance of the colony and SHIP was not line of sight they spent the nights in one of the Impervium tents for protection.

After eight months they had the new industries going well and the planet was manufacturing highly efficient electronics to use broadcast static electricity and some smaller jeep type vehicles that could use the electricity for motion. They used both petroleum and nuclear. They had developed a nuclear mining and processing system to fuel their defensive missiles that they now used for generating electricity. The advantage of nuclear was that it was safer than burning petroleum or coal and pollution free. It was hard for them to believe that the United States had developed peaceful nuclear power and used so little of it. France had used eighty percent nuclear for their electricity but then France had little coal and basically no petroleum within their country whereas the United States had plenty of everything. If it had not been for the enemy threat, they might have stuck with petroleum and coal for generations.

After nearly eight months they were ready to go on to Pyramid and help them catch up on technology. "Okay, who is going and who is staying. SHIP authorizes me and six of you. I want six because I still want to go back to the Library of Atlantis although we know it is much older than Atlantis."

Jill said, "I have been here on Pangaea for a long time. I'm going."

"Of course, I have missed you being on SHIP and we will need your knowledge both on Pyramid and if we go on to the library."

Estera said, "I am a permanent resident of Pangaea. I have four children now and am growing old like the rest of the colonists. I am still the youngest of my generation here."

"I didn't expect you to go. I know you only came with me because of your father and we made a deal. You were invaluable in your knowledge of botany and getting Earth plants established on our colony planets. I will see you whenever we get back here."

Dawn and Karen said, "We are permanent residents of SHIP and wherever it goes and you go, we go."

"Naturally, Sandy?"

"I stayed here before; I am not letting you go without me again."

"Shelby?"

"I chose to stay on Pyramid because it was more civilized and built up than Pangaea and I knew the difference. That did not turn out so well. I was too young for the rest of the colonists. I did find a good husband and a good family and then messed it all up. I lost my children and my health. SHIP gave me back my health. If it is okay with everyone, this time I think I want to be a permanent SHIP member and wife."

"As long as it is your decision. Adriana?"

"I am going. I was just another colonist on Pyramid when you recruited me and I like the adventure and the gathering of knowledge. I like being with you, Craig. I don't have the claim to you of Dawn, Karen, Jill and Sandy, but I belong with you, not on a planet, unless of course, you decide to stay on a planet."

"Jane?"

"Pangaea is so different than Vulcan. Not just the planet itself, but the people. I would like to just stay here like Estera. I have met other oriental colonists, and they are like I always wished the Chinese were. They are like the Chinese colony should have been. There are a lot of good Chinese, just not on Vulcan. I take that back. Some of the women on Vulcan could have been good, if they had the chance. Will you visit the women and when you come back this way let me know how they are doing?"

"Yes, but it will be over a year, maybe two years or more."

"That is okay. My age does not make as big a difference with the Orientals as it does with European races. There are many that look as young as I do. They may be in their mid to late thirties, but with some of them, they could be twenty-five. I was thirty-four when you rescued me from Vulcan. Some of the women on Vulcan were my age and looked fifty instead of thirty because of conditions. The women here were picked for their intelligence, genetics, and health and with the freedom and food here, they look as young as I do. Talking with SHIP I am really only twenty-two after SHIP fixed me, but mentally I am in my thirties. I can always pretend I am aging slowly because of my parents' genetics. I think I can find a husband and a family here and maybe have some children of my own in a free society here on Pangaea. Besides, six women is company and seven is too many for SHIP. My main contribution was electronic weapons. You may not even need the weapons and the beam for working with Impervium is fully developed both here and on Pyramid. The new efficient electronics from the library exceed my capabilities. By staying here, I can help bring that technology to a peak and contribute to the colony without having to worry about being forced into prostitution like on Vulcan. I want to stay here."

"Every one of you were great to me and accepted me as a member of your large family; the kind of family that all the colonies, including Vulcan was supposed to have. Craig, I do love you, but you have your maximum family for SHIP. If you had come up short, I would have consented to stay with you for as long as you needed me, but since your family is complete without me, I choose to stay here. I hope I will see all of you in the future. You will all have a special place in my heart."

Her speech brought tears to all including Jane, but Craig was able to reply, "You were critical to us. Without your weapons, Pyramid might have been destroyed and they might have found Pangaea too. Your discovery of the beam to work Impervium means we can work with Impervium as easy as mild steel. I'm sure you will be valuable here on Pangaea and I am glad that you like it and the people here."

"It is what the colony from China should have been. We might have looked further and found a new better planet than Vulcan or just integrated the Chinese into the other two colonies if they had been honest and created their colony ships as we directed. Yes, we will check in with Vulcan at some point in the near future and report back to let you know how the women are doing there."

"I will miss you personally, Jane, Hua Ming. I will ever be grateful that you held off helping Vulcan develop their weapons. I don't know if they could ever have conquered anyone, but they could have been dangerous with your help. I enjoyed our time together. I'm glad you like it here on Pangaea. You would think after being a world class gymnast you would have been treated better on Vulcan."

"I would probably not have been selected for the colony except for my fame as a gymnast. They considered me good breeding material. I'm glad I could escape before being subjected to the fate of the other women there. I will do well here and will be looking forward to seeing all of you again."

# CHAPTER TWENTY-EIGHT

# CONTACT

It took twelve weeks to get to Pyramid from Pangaea. Adriana was greeted by her old family that were now sixteen years older than she was thanks to SHIP taking her back to her prime age. Her old husband did not recognize her. She was wearing clothing designed by SHIP to match each person's inner desire in clothing. You could tell he was looking for her, but when in a group of young people in their twenties while everyone else was in their late thirties and early forties, they looked very young. Everyone on Pyramid were in their early twenties when they left Earth sixteen years ago putting everyone in either their late thirties or under fifteen years old, their children. Finally her old family recognized her because she had dark hair, was short, and it had to be her. Dawn had dark hair but was taller and not as busty and Karen had black hair and much taller. Jill had auburn hair and Sandy and Shelby had blond hair. Therefore the short young woman had to be Adriana. Her old husband was suddenly interested again.

"Adriana, is that you? You look so young and healthy."

Her old husband looked at her lustily her old husband's wives were pretty cool in greeting her. Probably because they were jealous of his obvious attraction to this young woman that only remotely looked like their old home mate.

"Yes, I know." She told the lie that they had invented to explain their ages instead of admitting SHIP had changed them. "Something about traveling at multiples of the speed of light. Einstein gave the theory that if you traveled away at near the speed of light and returned you might arrive before you left. That is not correct, but yes, I know that I look over ten years younger than when I left. I was still born thirty-seven years ago even if I look twenty-two or three.”

“But your scars are gone.”

“Good makeup. Makeup was a luxury in the colonies when I left, but the colony we came from had good makeup, so I brought some along.”

“Are you coming to the house to live again?”

“No. I am not. I was nearly an outcast when I got hurt, scarred and lost our child. I have a new family with Craig, the Captain. My sisters are Dawn, Karen, Sandy, Jill, and Shelby that tried living here until she too lost a child in the attack of the enemy ships. I belong with them and Craig now.”

“Do any of you have children?”

“No. No, children. We travel too much, too far, and too dangerous. Before we came back we followed the enemy ships after they broke off their attack here and followed them to their home planet where they were preparing a giant armada to totally annihilate life here. That is not something to do with children.”

“You mean the enemy is coming back in force?

“No, we annihilated their planet instead.”

“You are not serious.”

“Yes, I am. There were about a dozen ships escaped their planet before we wiped it clean.”

“How did you do that? What kind of weapons does your ship have?”

“Our ship has no weapons of its own. We arranged for their planet to run into a small moon, like Earth did, but worse.”

“But how?”

“Maybe someday, but right now I have to catch up with my current family and meet with your planet leaders.”

As Adriana caught up, Dawn asked her, “Well. Do you miss your old family?”

“Yes and no. My husband seems so old to me now. He was not interested in me when I had scars and had lost my ability to have children. Now, of course, he sees a young pretty woman whereas his wives are all pushing forty. It seemed like he was a dirty old man. His wives, that used to be my home mates, were pretty cold toward me. They saw my vibrant youth as a threat and hoped I would stay away.”

“I have been with you all and Craig long enough now that you are my real family. Basically, most of us were thrown together as colonists. It was not love that brought the families together with six wives per husband. It was more like choosing up sides for dodge ball. Finding the most compatible set to join with and not being too picky in case you might get stuck with the worst family. Considering, the family was pretty average and better than a lot of families. But my new family on SHIP is totally different.”

As they entered the conference center to meet with planet officials, Shelby was greeted by some old coworkers. She had been secretary of defense when she left on SHIP this last time. “Shelby, is that you? You are looking good. Are you here to stay?”

“No, I decided to stay with my original family. I thought I would like being a permanent planet resident, but now that I am back traveling space, I plan to stay with my original family.”

“We appreciate everything you did to develop our defenses before the enemy attacked. Are you back to help us build more defenses?”

“No, but we have some more new technology that may have multiple applications, but we will be here for months explaining.”

Craig interrupted, "Shelby. They know you here, would you please give the presentation on why we are here this time and what has happened since we left."

When the meeting adjourned Bill, Shelby's husband on Pyramid was there with her old family to greet her. "Shelby, you look so young and fit. When you left here you were looking older and had gained weight from the children. Now, you look just like when I first met you years ago. How did that happen?"

Shelby shook hands with Bill and her old housemates and gave them the pat answer about traveling at the speed of light for years. "No, I won't be coming back to your house. I did nothing wrong when I took my children to the daycare. It was not my fault that the enemy ships hit the daycare. You cast me out then, I think I am happy to be back with my original family." The same explanation she had given to her old co-workers.

.

Pyramid was not as pioneering as Pangaea and it took nearly a year to get the new technologies up to the same point that Pangaea had achieved in six months. The family found that since they could all teleport back and forth to SHIP alone anytime they wanted, they found it more comfortable to just spend their nights on SHIP. They spent many evenings visiting residents of Pyramid when they weren't working late. But whenever, one of them tired out, or finished a dinner at a friend's house, they teleported back to SHIP. This was now fifteen years since they left Earth when SHIP alerted everyone in the middle of the night.

"I have detected a small object that just dropped out of light speed six million miles from Pyramid and is decelerating toward Pyramid."

"Is it a weapon?" Karen asked.

"Unknown."

Dawn asked SHIP, "Have Pyramid defenses spotted it?"

"No. It is very small. Too small to carry a human."

"It has started transmitting a radio signal in English."

Craig asked, "Well?"

"I come in peace. This is a message capsule from the Galactic Alliance. It will come to a stop a million miles out so that you can come get it and scan it." Message repeats over and over.

"Okay, SHIP, show on all screens so we can watch it and keep your sensors searching for anything else suspicious. This could be a decoy or it could be an extremely powerful weapon that does not stop a million miles out."

Shelby contacted Pyramid Defense Command on video. The interior of SHIP was blanked out so only Shelby appeared on their screen while Shelby could see their background. "You recognize me as Shelby, your old secretary of defense. I have an emergency level message. Please bring your top duty officer to the screen so that we can talk."

In moments Shelby recognized one of the colonels of their defense team. "Hello Colonel Sams. I am no longer secretary of defense, but I have emergency information. Are you the highest ranking officer on duty?"

"Yes, I am, Shelby."

"Tell your defense forces to go on standby for a potential attack, but do not fire on anything unless you hear from me or lose communications from me."

"What is happening?"

"We are tracking a very small missile that is transmitting a message of peace. We do not know what it is, but it promises to stop one million miles from Pyramid and wait for our inspection. It could be a decoy ahead of some major attack or it could be what it says and is only a message for us. I don't want it shot down, but I want to make sure that you are ready for anything.

We will be taking our ship to intersect with it and inspect it. As you know, our ship is far superior to anything you have, but we have no offensive capability. Our sensors are superior to yours and our ship is infinitely safer from attack than yours. That is why we will inspect it, but you need to have your weapons at full readiness. You should call a full alert for the planet and bring in your top officials so we can keep them posted and have them available."

"Will do, Secretary Shelby."

"I'm not the secretary, but thank you for the complement."

Colonel Sams did as requested. All defensive forces were brought to full alert, the highest officials on the planet were brought into the command post with some remaining isolated in other shelters were linked by video to the command post. All manned space fighters were launched in widely spaced patrol patterns. All computer controlled defenses were brought to full readiness, but put on hold.

SHIP monitored the small missile closely while searching space to the limit of its sensors for anything else suspicious. The missile continued slowing and continued broadcasting its message on multiple channels of radio across the spectrums used on Earth. Obviously the builders of the missile had been listening to Earth broadcasts to know the frequencies and English language.

"Secretary Shelby, this is Pyramid Defense Secretary General Miller. We are now receiving the radio signals too. Is it interfering with this transmission?"

"No your transmission is clear here. Do you read me okay?"

"Yes."

SHIP said, "The missile stopped broadcasting on your video frequency shortly after you transmitted to Pyramid. It apparently heard my use of the frequency and stopped on that frequency.

Then as the Pyramid Command started using other frequencies to direct their forces, the missile quit on those frequencies too. The signal is not strong so our frequencies were overriding its signal anyway, but they cut their transmission on those frequencies. There is the normal delay of distance because it takes the radio signals several minutes to get here and several minutes for the missile to detect our use of frequencies. As it gets closer and reduces the number of frequencies its signals get stronger on the other frequencies. I cannot tell if it just has more power for transmission because of the reduced number of frequencies or because it wants its signals to be heard."

"Shelby, I hear a different voice in the background. It sounds a little like a computer generated voice in that it is flat without emotion."

"You are right, General. It is our ship's computer reporting using artificial intelligence, but it has no emotion."

In a few minutes, SHIP reported, "The missile appears to be almost stopped. It is not using retrorockets, it appears to be using antigravity like mine. This would indicate that it is not the old enemy, but some other society. It is also narrowing its broadcasts to frequencies used for routine broadcasts on Pyramid and increasing the power as it slows."

Karen said, "It wants everyone on Pyramid to hear the message. I have an issue though. If it had anti-gravity it could just maintain full past light speed and freeze to an instantaneous stop. Instead, it is slowing. Apparently they don't need retrorockets like ours, but something similar. Therefore it is not from Rogulatory, the planet that built SHIP."

Craig directed SHIP to approach the missile and intersect with its proposed stopping point while searching space for other objects. "SHIP, also look for rocks that are out of their old orbits in this solar system or its Ort Cloud and report anything."

"The only unexpected movement in the solar system is the unknown missile that is coming to a stop."

"Secretary Shelby. Everyone on Pyramid is receiving the broadcast now. Is it peaceful?"

"We will let you know shortly, General. It appears to be stopping where it said it would and we are moving to approach it for a full scan."

"Craig, it is requesting a direct link to do a data dump"

Adriana exclaimed, "Don't do it. It could act like a virus and take over the SHIP. It has advanced technologies, but as Karen said, it is not from Rogulatory or it would have antigravity drive and not need to spend time slowing."

"SHIP, you heard Adriana, don't link up with it. Can you safely transmit that decision without linking up?"

"Yes."

Adriana asked SHIP, "Can you quick make up a robot ship with a computer memory that could link with the foreign missile that you could then bring back on board and have us humans communicate directly with the robot instead of you linking with the missile?"

"Yes. Craig?"

"Do it."

SHIP reported, "I see no hint of offensive weaponry. There is no hint of radiation. All transmissions have stopped as if it is awaiting inspection."

"Okay, SHIP, send over the robot ship."

"The missile is transmitting extreme high speed data to the robot ship. It started out too fast for the robot to receive it, but has slowed down to the maximum transmission speed for the robot to record one hundred percent with error checks for accuracy.

The robot reports that it is complete. I will bring it back on board for your inspection if you agree."

"Go ahead, but keep it away from the control room, just in case." The small robot ship appeared in the hanger where the one shuttle was kept.

"I am going into the hanger first. I don't want anyone else until I think it is safe." Karen declared.

Karen got her way and then said, "SHIP has warmed it to be SHIP temperature and it appears safe."

Adriana said, "I want to be next as I am the computer expert."

"Okay, but I am coming with you." said Craig. "SHIP, can you direct the robot to communicate with us in English and if there is video to one of our tablet computers without transmitting to your memory or any other computer?"

"Will do."

"Dawn, Jill, come on in. Sandy, Shelby, stay in the control room and Shelby, give a report to Pyramid Command."

"General. We could not find any dangers from the missile which is totally stopped. We sent out a robot ship computer to link with it and download information. We have our robot ship back on board and are getting ready to see what we downloaded from the missile."

"Is it safe?"

"We think so. The missile requested link with the ship's computer, but we did not allow that which is why we sent the robot computer to do the link. We will let you know what we find out."

The message from the missile that had been downloaded to the robot computer ship that SHIP has sent out to link with the missile said, "Welcome to the Galactic Alliance. We have learned your language from years of monitoring your radio and television transmissions. We have detected your escape from the third planet, that you called Earth. We saw your detonation of nuclear weapons and were concerned about your society. We detected your access to the pyramid on the planet you named, Pyramid. We detected your access to the repository on the planet you called Earth. We detected the attacks on your society from the peoples you called the enemy. We detected your destruction of what you called the enemy planet. We do not know what weapon you used."

"As you know, electronic transmissions are limited to the speed of light, so we sent this message by physical ship. It detected your ship has anti-gravity and therefore is very advanced. We have partially accepted the people you called Martians after they escaped from Earth. They have been shown to be peaceful, but they could not defend themselves from the Martians you called your enemy. The Martians did not have antigravity ships like yours. Therefore they are not full members of the Galactic Alliance. Your possession of antigravity ships and weapons that can destroy a planet means that you are advanced enough that we want to have a treaty of understanding with you. This treaty will make your people full members of the Galactic Alliance, but only if you agree to turn over your science to us and agree not to use your science to attack any members of the Galactic Alliance."

"Because of your connections to the people you call Martians we will tell you about them and where to find their planets if you do not know. The Martians made a very bad error on two of their planets. They tried to use their power pyramids to transmit an electronic signal faster than light. The feedback between the two planets destroyed most of life between those planets. There were only a few survivors that were eventually evacuated to other Martian planets leaving those two planets mostly destroyed."

"If you agree to this treaty, send your recorded answer back to the coordinates given. These coordinates will not lead you to any of our planets, but will be monitored and your answer received. If your message missile has the same speed it will be received at the receiving location within one rotation of your planet around the sun you orbit."

"If your reply is accepted you will receive our acceptance within one additional of your years."

The interstellar locations of the Martian planets and the receiving location for the Galactic Alliance was transmitted as code that only Dawn and SHIP could make sense of. Dawn said, "I think I can understand these locations, but it may take me months even with SHIP's computer. I just don't know. I think my theories should be universal ideas of space navigation, but I do not know their way of thinking or what they may consider space landmarks to guide from."

"Do the best you can, Dawn." Craig went on to say, "Shelby, relay this information to Pyramid Command. We will want to talk about what to do next. Everyone, I do not know what to make of this. Are they peaceful? Do they just want our technology? They don't have antigravity drive like SHIP's which makes them possibly more advanced than we are or maybe not as advanced. They know where we are, but we don't know where they are. Presumably they must be close to the location they want us to send the answer to or they would not promise to answer back in one year. We have SHIP, but SHIP does not know how its antigravity works to provide it to us."

Karen said, "I think we should give them a courteous reply, but don't tell them anything about us and then go find us some Martians and try to make peaceful contact with them. Maybe they will know more about this Galactic Alliance. The Martians have been out in space fifty thousand more years than we have. Whatever we do, we don't want to admit our weakness or give them the idea of pushing a rock into a planet. I am thinking that the bad Martians the repository told us about had nothing to do with pushing a rock into Mars or Earth. It could have been pure chance and maybe we are the only ones in the galaxy that actually did it."

Dawn agreed, "I agree completely with Karen. We don't know if they are good guys or bad guys."

Craig hit the nail on the head, "I agree, but whoever this Galactic Alliance is they are not the ones that built SHIP or the repository on Earth. They detected the antigravity of SHIP but knew nothing about us as people or the mission of SHIP and the repository or repositories if the repositories are all around the galaxy. Therefore we know that there is us, the supposedly good Martians, the bad Martians that escaped their planet, this Galactic Alliance, and then there is the intelligence of Rogulatory that built SHIP and the repository. We can be pretty sure that whoever built the repository and SHIP were good guys, but this Galactic Alliance may be their enemy."

They all agreed. Shelby suggested that they consult with Pyramid Command for their ideas. At least they knew that there was other intelligent life out there besides themselves and the Martians.

Karen said, "With possibly of between eleven and thirty billion habitable planets just in the Milky Way Galaxy there could be millions of planets with intelligent life. Not all of which will be friendly to the new guy on the block."

Craig declared, "We need to send a reply within the next few weeks, whatever it is, and then we have at least two years to find the Martians and possibly some trace of the Galactic Alliance. The Galactic Alliance will not know how long we spent building a reply and a message missile to carry it back."

"Jill I want you to be working on such a ship with another ship inside that can report back to us when the larger ship is approached in space. Shelby, do you think there is any way to build a tracer into the message ship that can track them? Adriana, can you build in some sort of virus that can get information from them to send back to us by the smaller message ship without them knowing we did it? SHIP, if we were to go to the point in space that we are directing the message ship, could you hide from them?”

# The Crew

Craig Decker – Captain

Dawn Everly – astronavigation

Jill Leary – Petrochemical engineering and Geology.

Shelby Denton– Cheerleader – Chemistry

Estera Shea— Skidmore – Botany

Sandy Scofield – beach –Zoology

Karen Martin aka She Who Wanders - Survivalist

Hua Ming (Jane) – Electrical Engineering Chinese gymnast from Vulcan replaced Estera who stayed on Pangaea

Adriana DeCamp – Computer Science from Pyramid replaced Shelby who stayed on Pyramid

# ABOUT THE AUTHOR

The Author spent over 39 years with the United States Air Force in the USA, Asia, and Europe. He had some part in the development of most of the new weapons system in development between 1979 and 1989. Some of those systems are just now being delivered. He taught classes in weapons system research and development and logistics management as a guest lecturer at the Air Force Institute of Technology and various conference rooms around the USA. Over 3000 current and future managers attended his class on R&D scheduling. He served on many brain storming teams to come up with unique solutions to military problems from shooting down space objects to moving “dud” bombs off an active runway.

He is now retired from both active duty and civil service. He is still active in writing, the local amateur radio club, towing the Shriner float in parades and going to almost weekly dances and generally enjoying himself.

www.ingramcontent.com/pod-product-compliance
Lightning Source LLC
LaVergne TN
LVHW050909210926
848482LV00012B/379
* 9 7 9 8 8 6 9 1 6 2 3 4 2 *